MORAL POLITICS

MORAL POLITICS

HOW LIBERALS AND CONSERVATIVES THINK

Third Edition

GEORGE LAKOFF

THE UNIVERSITY OF CHICAGO PRESS
Chicago and London

The University of Chicago Press, Chicago 60637
The University of Chicago Press, Ltd., London
© 1996, 2002, 2016 by George Lakoff
All rights reserved. Published 2016.
Printed in the United States of America

25 24 23 22 21 20 19 18 17 16   2 3 4 5

ISBN-13: 978-0-226-41129-3 (paper)
ISBN-13: 978-0-226-41132-3 (e-book)
DOI: 10.7208/chicago/9780226411323.001.0001

Library of Congress Cataloging-in-Publication Data

Names: Lakoff, George, author.
Title: Moral politics : how liberals and conservatives think / George Lakoff.
Description: Third edition. | Chicago : The University of Chicago Press, 2016. |
    Includes bibliographical references and index.
Identifiers: LCCN 2016005827 | ISBN 9780226411293 (pbk. : alk. paper) |
    ISBN 9780226411323 (e-book)
Subjects: LCSH: United States—Moral conditions. | Political ethics—United States. |
    Social ethics—United States. | Social values—United States. | Conservatism—
    United States. | Liberalism—United States.
Classification: LCC HN90.M6 L34 2016 | DDC 306.0973—dc23 LC record available
    at http://lccn.loc.gov/2016005827

To My Wife
KATHLEEN FRUMKIN

and to the Memory of
PAUL BAUM

Contents

Part IV: The Hard Issues

Part V: Summing Up

Part VI: Who's Right? And How Can You Tell?

Preface to the Third Edition

Moral Politics is even more relevant today than it was when its first edition appeared in 1996. Conservatives and progressives have opposing worldviews at the level of basic morality—they have conflicting understandings of what is right and wrong. The social and political divide in the U.S. is even stronger now than it was two decades ago. Accordingly, the need for a public understanding of the nature of that divide has grown enormously. And this is not only the case in the U.S. Similar divides have been fostered in many other countries, often supported by powerful conservative groups in the U.S.

The fundamental nature of the divide has not changed, although the world has changed enormously. The computer revolution has made cellphones and laptops—along with internet access and social media—ubiquitous. The effects of global warming are also ubiquitous: massive storms and floods, droughts and fires, sea-level rise and water-shortages, mass migrations of fish and birds, as well as species extinction. Scary things have happened: the 9/11 attack; wars in Iraq and Afghanistan; the Syrian government's horrific attacks on its own people with the resulting flood of refugees throughout Europe and the rise of the horrific ISIL regime; the rise of China as an economic and military world power buying up land throughout the world; the rise of Putin and his rebuilding of the Russian military; the economic crash of 2008 and the ever-increasing wealth of the 1 percent along with the demise of the American middle class; and the Citizens United decision by the conservative Supreme Court pumping billions into the Republican election coffers and conservative causes at all levels.

One might have hoped that such massive changes might have brought people together, but the reverse has happened. The divisions between conservative and progressive values have become stronger and more virulent. They cry out for a public understanding. Hence, this new edition of *Moral Politics*.

What follows is the text of the second edition followed by a new afterword to bring it up to date. But some of what has been learned over two decades should be stated at the outset.

All Politics Is Moral

When a political leader proposes a policy, he or she assumes that the policy is right, not wrong or morally irrelevant.

Conservatives and progressives typically support opposite policies because they have opposite moral worldviews—opposite notions of right and wrong. Moral worldviews are important to people, part of their self-identity. People tend to think of themselves as good and moral, not taking into account that there could be an opposite view of what is moral.

Lists of Apparently Unrelated Issues

Progressives and conservatives often identify themselves via lists of positions on issues which they are for or against: legalized abortion; higher taxes on the rich; stopping and reversing global warming; increases in Social Security, Medicare, and other social safety nets; gun control; raising the minimum wage; gay marriage; equal pay for women; providing undocumented immigrants and their children with health care and public education; universal early childhood education; support for unions; and so on. There are scores of such issues. And, as we shall see, one's positions on them are not random or arbitrary; they generally follow from one's moral worldview. The bulk of this book shows how.

Conservatives usually have a smaller list of apparently unrelated general principles. They vary somewhat, but here is a typical short list: smaller government; free markets; personal responsibility; lower taxes; strong defense; traditional family values. Many conservative positions follow from these principles. But they are still a list and do not mention the general moral basis of conservative thought from which they all flow: Strict Father Morality.

What Is a "Moderate?"

A moderate conservative has mostly conservative views but progressive views on some issues that may vary from person to person. A moderate progressive is mostly progressive but has some conservative views on some issues.

Moderates of this kind are called "bi-conceptuals," since they have two opposing worldviews at once applied to different issues, one worldview more strongly held or more widely applicable than the other.

Alternatively, a moderate progressive or conservative may be "pragmatic," that is, willing to compromise to get as close as possible to his or her strongly held beliefs.

There Is No "Middle"

The metaphor of the Left versus the Right comes with a line between them, and the metaphorical line has a "middle." There are "moderates" like those above but not one worldview of the Moderate shared by all moderates. In short, there is no single "middle" defined by single worldview.

Neural Politics

All thought is physical, carried out by neural circuitry. No thought just floats in midair. Research over the past four decades has provided insight into how neural circuitry carries out thought that is below the level of consciousness.

Most Thought Is Unconscious

Only a tiny amount of our thought is conscious. A typical estimate is about 2 percent, with about 98 percent of thought unconscious.

Moral worldviews, like most deep ways of understanding the world, are typically unconscious. The more that a neural "idea-circuit" is used, the stronger it gets—and may eventually become permanent, effectively "hard-wired." Hence, most of what we will be discussing in this book occurs at the neural level and is likely to be unconscious.

Unconscious thought is studied in the field of cognitive science.

What If the Facts Don't Fit the Worldview?

We know from experiments that conscious perception is not immediate. To recognize a visual input, a sound, or a touch input you have to have in your brain neural circuitry that is able to recognize it—that fits it. What if your sense input doesn't fit what is in your brain? Your brain changes it, if possible, to make it fit.

Inputs from the senses presented to the eyes, ears, or touch take about one hundred milliseconds (a tenth of a second) before they can become conscious. That is so fast for conscious functioning that we don't notice any difference between the input to the senses and our conscious perception. But neurons fire on the scale of one millisecond (one thousandth of a second), and it take three to five milliseconds to fire again. It takes many neurons and a sequence of neural firings to take a sense input and turn it into a conscious perception. In that time, it is common for the visual, sound, or touch system to make a change, cancelling out part of what is present to the senses and creating a new input that fits the circuitry already in your brain.

This does not just happen in experiments with flashes of light, beeps, fast touches on the arm, and pictures and sounds of someone pronouncing syllables. It also happens with facts presented in language. If the facts don't fit one's worldview, one of several things can happen:

- The fact may be changed to fit your worldview.
- The fact may be ignored.
- The fact may be rejected and possibly ridiculed.
- Or the facts, if threatening to your worldview, may be attacked.

All of these happen in political discourse. Deep and persisting moral worldviews tend to be part of your brain circuitry and tend to become part of your identity. In most cases, the neural wiring—and your identity—stay, and the facts are ignored, dismissed, ridiculed, or attacked. It takes extraordinary openness, training, and awareness of this phenomenon to pay critical attention to the vast number of facts we are presented with each day. Few members of the general public—or those in politics or the media—fit this profile.

This isn't always so. Some facts are so traumatic that they effect a deep change. Such a change happened to me when I saw the image of the Twin Towers falling on 9/11—and when I heard of the assassinations of Martin Luther King Jr. and John and Robert Kennedy. They also happened to me in the early 1970s when I read the earliest report that the temperature of the earth's atmosphere had recently risen one degree centigrade. Recalling my MIT thermodynamics course, I had a traumatic realization: That is a huge amount of heat that would have massive horrible effects! But one degree sounds small and most people dismissed the report or never seriously entertained that fact, if they even noticed it. We now know why.

The normal dismissal of daily facts that don't fit moral worldviews explains why so many conservatives deny global warming in the face not only of the vast range of scientific facts, but even in the face of images of melting glaciers and the reality of droughts and fires. It is not that science deniers say to themselves, "I'm going to deny the scientific facts." Instead, their brains work automatically and unconsciously so as to produce the effect of science denial.

But science denial is not just relegated to conservatives. Many liberals took courses in fields like political science, economics, public policy, and law. They implicitly learned a worldview about reason itself, a worldview that is at odds with the scientific facts from the cognitive and brain sciences. They learned a centuries old theory of rationality that says that thought is conscious (when it is mostly unconscious), that it works by logic (it actually works by embodied primitives, frames, conceptual metaphor, and conceptual integration), that all people have the same logic (which is supposed to be what makes us *rational* animals). As a consequence, it should be true that if you just get the facts out to people, they will reason to the right conclusion. And so year after year, decade after decade, liberals keep telling facts to conservative audiences without changing many minds. This behavior by liberals is itself a form of science denial—the denial of the cognitive and brain sciences. It is simply *irrational* behavior by many people proud of their rationality.

It is for this reason that so many liberals have a low opinion of conservatives, considering them to be either uninformed, stupid, greedy, mean, or just nuts. Some may be, just as some liberals may be. But on the whole, conservatives are normal people who happen to have a conservative moral worldview deeply embedded in their brains and whose personal identity is significantly defined by that worldview.

It is not that facts don't matter. They obviously do—enormously. But the facts have to be framed in appropriately moral terms so that they can be taken seriously. To do that, you have to understand the worldviews of the people you are talking to, whether you are a liberal or a conservative. You have to know whether or not they are hardcore believers or bi-conceptual or pragmatic moderates. To have any hope of healing the divisions in our culture, we need to understand the worldview problem and make it part of public discourse.

It is this hope that motivates this third edition of *Moral Politics*.[1]

George Lakoff
Berkeley, CA, November 2015

---

1. For recent survey-based and experimental results confirming the analysis given in this book, see Wehling, E. 2013. "A nation under joint custody: How conflicting family models divide US-politics. PhD diss., University of California at Berkeley.

—

Acknowledgments

I would like to begin by expressing my gratitude to my students and colleagues at the University of California at Berkeley, where I have had the privilege of teaching for the past twenty-four years. The extraordinary faculty and the community of brilliant, dedicated, gentle, and yet constantly questioning students have provided me with an environment of incomparable nurturance and intellectual stimulation, for which I am deeply grateful. I would especially like to thank Noah Baum, Diana Baumrind, Robert Bellah, Joe Campos, David Collier, Joe Grady, Joshua Gutwill, J. B. Lowe, Pamela Morgan, Paul Mussen, Srini Narayanan, Alan Schwartz, Laura Stoker, Eve Sweetser, Sarah Taub, Eliot Turiel, Nancy Urban, and Lionel Wee. Pamela Morgan also provided invaluable bibliographical help.

Mark Johnson's *Moral Imagination* inspired me to start thinking about moral systems, and Mark has talked me through much of this book. I cannot imagine a more creative colleague or a better friend.

My son, Andrew Lakoff, read through more drafts than filial duty could possibly have required. The book is far better for his criticisms, challenges, good humor, excellent advice, and productive suggestions.

Michele Emanatian, Zoltán Kövecses, Jim McCawley, Mark Turner, and Steven Winter provided important comments and, as usual, gave me a lot to think about, as did Antonio and Hannah Damasio.

I am extremely fortunate to have as friends and colleagues Paul Deane and David Tuggy who, despite their political disagreements with me, offered me a great deal of their insight and wisdom. They won't agree with a great deal of this book, but their patient and kind discussions have made it a better book than it would otherwise have been.

Geoff Huck, the linguistics editor at the University of Chicago Press, provided invaluable suggestions for improving the manuscript. I am grateful for his rare combination of good editorial judgment, a fine ear, and a superb linguist's mind. Several anonymous reviewers for the Press provided very helpful suggestions for revision.

In the midst of working on this book, I had the opportunity to visit and teach at the École des Hautes Études En Sciences Sociales in Paris and at the University of New Mexico 1995 Summer Linguistics Institute. I would especially like to thank Pierre Encrevé, Michel de Fornel, and their students at the École for lengthy informative discussions of these topics during my lectures there, as well as Pierre Bourdieu, who made stimulating suggestions. I am grateful to Joan Bybee, director of the Linguistics Institute, for making it possible for me to teach there, and to Sherman Wilcox, who made it possible for me to present some of this material at the 1995 meeting of the International Cognitive Linguistics Association. I would also like to thank Paula Bramante of UNM for her suggestions.

I am grateful to the University of California for a President's Humanities Fellowship during 1994–95.

Cafe Fanny and O Chame in Berkeley provided superb sustenance.

My wife, Kathleen Frumkin, talked through this with me and helped in many, many other ways as well.

This book began with a conversation in my garden several years ago with my friend the late Paul Baum. I asked Paul if he could think of a single question, the answer to which would be the best indicator of liberal vs. conservative political attitudes. His response: "If your baby cries at night, do you pick him up?" The attempt to understand his answer led to this book.

Part One

Introduction

— 1 —

The Mind and Politics

Contemporary American politics is about worldview. Conservatives simply see the world differently than do liberals, and both often have a difficult time understanding accurately what the other's worldview is. As a student of the mind and of language, I think we can make much better sense than has been made of the worldviews and forms of discourse of conservatives and liberals.

I work in a discipline that studies how people conceptualize the world. It is called cognitive science, which is the interdisciplinary study of the mind. It is a broad discipline, covering everything from vision, memory, and attention to everyday reasoning and language. The subfield most concerned with issues of worldview, that is, with everyday conceptualization, reasoning, and language, is cognitive linguistics. I have been a cognitive linguist virtually from the birth of the field, and it is my profession to study how we conceptualize our everyday lives and how we think and talk about them. The study of political concepts and political discourse falls under the job description of those in my field, though until now research in the area has been relatively sparse.

Common Sense and Unconscious Thought

A few words about my profession might be useful at the outset. One of the things most studied in cognitive science is common sense. Common sense cannot be taken for granted as a given. Whenever a cognitive scientist hears the words "It's just common sense," his ears perk up and he knows there's something to be studied in detail and depth— something that needs to be understood. Nothing is "just" common sense. Common sense has a conceptual structure that is usually unconscious. That's what makes it "common sense." It is the commonsensical quality of political discourse that makes it imperative that we study it. I hope that you will see by the end of this book just how deep, complex, sophisticated, and subtle common sense is, especially in the domains of morality and politics.

One of the most fundamental results in cognitive science, one that comes from the study of commonsense reasoning, is that most of our thought is unconscious—not unconscious in the Freudian sense of being repressed, but unconscious simply in that we are not aware of it. We think and talk at too fast a rate and at too deep a level to have conscious awareness and control over everything we think and say. We are even less conscious of the components of thoughts— concepts. When we think, we use an elaborate system of concepts, but we are not usually aware of just what those concepts are like and how they fit together into a system.

That is what I study: what, exactly, our unconscious system of concepts is and how we think and talk using that system of concepts. In recent years, my work has centered on two components of conceptual systems: conceptual metaphors and categories, especially radial categories and prototypes. A conceptual metaphor is a conventional way of conceptualizing one domain of experience in terms of another, often unconsciously. For example, many people may not be aware that we commonly conceptualize morality in terms of financial transactions and accounting. If you do me a big favor, I will be *indebted* to you, I will *owe* you one, and I will be concerned about *repaying* the favor. We not only talk about morality in terms of paying debts, but we also think about morality that way. Concepts like retribution, restitution, revenge, and justice are typically understood in such financial terms. As we shall see, examples like these are the tip of the iceberg. Much of moral reasoning is metaphorical reasoning, as will become apparent below.

It should also become apparent, if this example does not already make it clear, that metaphorical thought need not be poetic or especially rhetorical. It is normal, everyday thought. Not every common concept is metaphorical, but a surprising number are (see References, sec. A1).

Metaphorical Common Sense

Much of what we read on the daily op-ed pages of our finest newspapers is metaphorical commonsense reasoning. Let us consider a very simple example, taken from a column by *Washington Post* columnist William Raspberry (as it appeared in the *Houston Chronicle,* section A, p. 30, February 4, 1995). The column begins straightforwardly enough.

> The government of the District of Columbia is reeling from a newly discovered budget shortfall of at least $722 million and there is growing talk of a congressional takeover of the city.

After an example of spending he considers questionable, Raspberry says,

> What is about to do us in . . . is the poor but compassionate mother with a credit card.
> To put it another way, a huge amount of the city's

stupendous debt is the result of the local government's effort to do good things it can't afford.

He then gives a list of examples of good things the city
government wants to do and which he thinks it can't afford,
and finishes the column as follows:

> But a good chunk of the underlying problem is the
> compassionate mom attitude that says: If it's good for
> the kid to have, then I ought to buy it—and worry
> later about where the money will come from.
>
> Well, Mom not only has reached her credit limit:
> she's in so much trouble that scrimping and saving
> now won't solve the problem. She'll need a bailout
> from Congress.
>
> But then, she has to learn to say no—not just to
> junk food but to quality cuts of meat she can't afford.

None of Raspberry's readers have any problem understanding this column. He writes it as if it were just common
sense. Yet, it is an elaborate conceptual metaphor, and he is
reasoning in terms of this metaphor.

In the metaphor, the government is an overindulgent, impractical mother and the citizens are her children. She has
no self-discipline; she is indulging her children irresponsibly,
using money she doesn't have. This is not merely politics,
it is a story with a moral. The moral is that Mom will have
to learn self-discipline ("to say no") and self-denial ("to
quality cuts of meat she can't afford"). Only then will she
be a good mother.

We all understand this column, and to many readers it
will seem like common sense. But why? Is the metaphor that
government is a parent and the citizens are children newly
made up? Or is it familiar, a metaphor we already know?
And why should readers be willing to reason about a government in this way? Why don't they just reject the metaphor

as ridiculous? Why don't readers—all readers—say in response, "What's all this nonsense about indulgent moms?
Let's get real and talk about the details of economics and
policy." But readers don't. The column is "just common
sense." And moreover, it is conservative common sense.
The logical structure of the column is determined by metaphor, not by facts. One could have taken the same budget
shortfall and framed it in a different way. One could have
observed that Washington, D.C., must have city services
beyond its population to serve the large number of relatively
well-off civil service workers, lobbyists, and others who live
in the wealthy suburbs but work in town. One could also
have mentioned that it is the *responsibility* of Congress to
see that the city is maintained properly and that it lives by a
humane standard, indeed that it should set a standard for the
country. One could then apply the metaphor of the government as parent to Congress, seeing Congress as a deadbeat
dad, refusing to pay for the support of his children, the citizens of Washington, D.C. One could then have drawn the
moral that deadbeat dad Congress must meet his responsibilities and pay, no matter how tough it is for him. This is just
common sense—a different kind of common sense.

What, exactly, is conservative common sense? How does
it differ from liberal common sense? And what role, exactly,
does metaphorical thought play in the everyday commonsense reasoning of conservatives and liberals? As we shall
see, the metaphor used in this column, that of the government as parent, has a great deal to do with conservative
common sense in general, as well as with what conservatism
is as a political and moral philosophy.

RADIAL CATEGORIES

Radial categories are the most common of human conceptual
categories. They are not definable in terms of some list of

properties shared by every member of the category. Instead, they are characterized by variations on a central model. Take the category *mother*. The central model is characterized by four submodels. (1) The birth model: the mother is one who gives birth. (2) The genetic model: the mother is the female from whom you get half your genetic traits. (3) The nurturance model: your mother is the person who raises and nurtures you. And (4) The marriage model: your mother is the wife of your father. In the most basic case, all conditions hold. But modern life is complex, and the category extends to cases where only some of these conditions are met. Hence, there are special terms like birth mother, genetic mother, foster mother, stepmother, surrogate mother, adoptive mother, and so on.

Another example of a radial category is forms of harm. The central case is physical harm. But the category also includes kinds of harm that are metaphorically understood in terms of physical harm, e.g., financial harm, political harm, social harm, and psychological harm. Our courts recognize that these are all forms of harm, yet they also recognize the centrality of physical harm, for which the most severe penalties are usually reserved.

Radial categories, with central cases and variations on them, are normal in the human mind. And, as we shall see, the categories of conservative and liberal are also radial categories. This is important to realize because conservative and liberal are very complex categories, with a great many variations. The theory of radial categories allows us to account for both the central tendencies and the variations. For an introduction to radial categories, see Lakoff (1987) and References, sec. A2.

TYPES OF PROTOTYPES

The central members of radial categories are one subtype of a general phenomenon called "prototypes" (see References, sec. A2). There are many types of prototypes and it is important to discuss them at the outset since they will play a major role throughout the book. A prototype is an element of a category (either a subcategory or an individual member) that is used to represent the category as a whole in some sort of reasoning. *All prototypes are cognitive constructions used to perform a certain kind of reasoning; they are not objective features of the world.*

Here are some of the basic types of prototypes that play a role in American politics and that will recur throughout this book:

1. The *central subcategory* of a radial category: This provides the basis for extending the category in new ways and for defining variations.

> Political examples will include central types of liberals and conservatives.

2. A *typical case* prototype: This characterizes typical cases and is used to draw inferences about category members as a whole, unless it is made clear that we are operating with a nontypical case.

> For example, what we consider to be typical birds fly, sing, are not predators, and are about the size of a robin or sparrow. If I say "There's a bird on the porch," you will draw the conclusion that it is a typical case prototype, unless I indicate otherwise. If I speak of a typical American, what comes to mind for many is an adult white male Protestant, who is native-born, speaks English natively, and so on.

3. An *ideal case* prototype: This defines a standard against which other subcategories are measured.

> We will be discussing what conservatives and liberals think of as an ideal parent, an ideal citizen, and an ideal person.

4. An *anti-ideal* prototype: This subcategory exemplifies the worst kind of subcategory, a "demon" subcategory. It defines a negative standard.

> Liberals and conservatives have very different kinds of demons, and we will discuss the types and how they are used in reasoning.

5. A social *stereotype:* This is a model, widespread in a culture, for making snap judgments—judgments without reflective thought—about an entire category, by virtue of suggesting that the stereotype is the typical case.

> Social stereotypes are commonly used in unreflective or biased discourse. Examples include the Drunken Irishman (used to suggest that the Irish typically drink to excess), the Industrious Japanese (used to suggest that the Japanese are typically industrious), and so on. Ethnic and gender stereotypes constantly enter into political discourse, as do political stereotypes. Stereotypes can either be based on myth or on individual well-known examples.

6. A *salient exemplar:* A single memorable example that is commonly used in making probability judgments or in drawing conclusions about what is typical of category members.

> It is commonplace in political discourse to use a salient exemplar as if it were a typical case; for example, popularizing the case of a single welfare cheater to suggest that everyone on welfare cheats.

7. An *essential* prototype: This is a hypothesized collection of properties that, according to a commonplace folk theory, characterizes what makes a thing the kind of thing it is, or what makes a person the kind of person he is.

Essential properties of birds are, among others, that
they have feathers, wings, and beaks, and lay eggs.
Rational thought is seen as an essential property of human beings. In moral discourse, the notion of *character* is seen as defined by an essential prototype. Your
character is what makes you what you are and determines how you will behave.

None of these should be strange or unfamiliar. All of these
are normal products of the human mind, and they are used
in normal everyday discourse. There is nothing surprising
about their use in politics, but we do need to be aware of
how they are used. It is important, as we shall see, not to
confuse a salient exemplar with a typical case, or a typical
case (say, the typical politician) with an ideal case (like the
ideal politician).

The Present Book

In my previous writings I have been concerned with the
details of conceptual analysis and their consequences for
such fields as cognitive science, philosophy, and linguistics.
The present book came out of such routine research. Around
the time of the conservatives' victory in the 1994 elections,
I happened to be working on the details of our moral conceptual system, especially our system of metaphors for morality.
During the election campaign, it became clear to me that
liberals and conservatives have very different moral systems,
and that much of the political discourse of conservatives and
liberals derives from their moral systems. I found that, using
analytic techniques from cognitive linguistics, I could describe the moral systems of both conservatives and liberals
in considerable detail, and could list the metaphors for morality that conservatives and liberals seemed to prefer. What
was particularly interesting was that they seemed to use virtu-

ally the same metaphors for morality but with different—almost opposite—priorities. This seemed to explain why liberals and conservatives could seem to be talking about the same thing and yet reach opposite conclusions—and why they could seem to be talking past each other with little understanding much of the time.

At this point, I asked myself a question whose answer was not at first obvious: What unifies each of the lists of moral priorities? Is there some more general idea that leads conservatives to choose one set of metaphorical priorities for reasoning about morality, and liberals another? Once the question was posed, the answer came quickly. It was what conservatives were talking about nonstop: the family. Deeply embedded in conservative and liberal politics are different models of the family. Conservatism, as we shall see, is based on a Strict Father model, while liberalism is centered around a Nurturant Parent model. These two models of the family give rise to different moral systems and different discourse forms, that is, different choices of words and different modes of reasoning.

Once we notice this, a deeper question arises: Can we explain what unifies the collections of liberal and conservative political positions? Do models of the family and family-based moral systems allow one to explain why liberals and conservatives take the stands they do on particular issues? The problem is a difficult one. Consider conservatism. What does opposition to abortion have to do with opposition to environmentalism? What does either have to do with opposition to affirmative action or gun control or the minimum wage? A model of the conservative mind ought to answer these questions, just as a model of the liberal mind ought to explain why liberals tend to have the cluster of opposing political stands. The question of explanation is paramount. How, precisely, can one explain why conservatives and liberals have the clusters of policies they have?

Even the basics of conservative and liberal discourse require explanation. Conservatives argue that social safety nets are immoral because they work against self-discipline and responsibility. Liberals argue that tax cuts for the wealthy are immoral because they help people who don't need help and don't help people who do need help. What moral systems lead each of them to make these arguments and to reject the arguments of the other? Why do conservatives like to talk about discipline and toughness, while liberals like to talk about need and help? Why do liberals like to talk about social causes, while conservatives don't?

The answers, I will argue, come from differences in their models of the family and in family-based morality—from the distinction between what I will call the Strict Father and Nurturant Parent models. The link between family-based morality and politics comes from one of the most common ways we have of conceptualizing what a nation is, namely, as a family. It is the common, unconscious, and automatic metaphor of the Nation-as-Family that produces contemporary conservatism from Strict Father morality and contemporary liberalism from Nurturant Parent morality.

This is not something that is either straightforward or obvious, partly because the two moral systems are not obvious. But once we discern the moral systems, the explanation emerges. It is the only explanation there is at present for why conservatives and liberals have the policies they have and use the modes of reasoning and the language that they use.

VARIATIONS

There are, of course, far more than two forms of morality and politics, even among conservatives and liberals. But set within normal human minds, the two family systems and moral systems that I will be outlining give rise, in a systematic way, to a considerable number of actual moral and political positions, each of which is a variation on one of the two systems. Such variations occur systematically because human category-structure is radial. A given central model, by virtue of its structure, allows a certain range of variations, but not just any variations. The parameters of variation are defined by the structure of the model, as we shall see in detail below. As a result, two conservatives such as Robert Dole and Phil Gramm may have very different positions on a certain range of issues yet still be conservatives. One goal of this study is to provide a theory of what determines such parameters of variation. Thus, when I speak of "two models," I will be referring to two central models whose structures determine a wide range of variations. Thus, "the" model of liberalism is the central model whose structure naturally gives rise to a wide variety of liberalisms. What makes all these liberalisms a single category is the systematic relationships they bear to the central model. The details of the parameters of variation that lead to such complex radial categories will be discussed in Chapters 5, 6, and 17 below.

COHERENCE

What I will call "central" conservatism and liberalism are coherent political ideologies. Moreover, each variation of conservatism and liberalism is a coherent ideology. The radial categories show how the coherent ideologies in each category fit together and what the relationships among them are.

But not all citizens have coherent ideologies. Far from it. Indeed, one of the important results of the study of conceptual systems is that they are not internally consistent. It is normal for people to operate with multiple models in various domains. Thus, one may have a number of inconsistent models of what a marriage should be like or how a computer works. Sometimes one model—one precise model—is used, sometimes another precise model is used. If one fails to look at the reasoning used in each case, it might appear that there are no models at all, that people just act randomly. But a look at forms of reasoning used on a case-by-case basis reveals that, for the most part, different models are being used in different instances.

One of the things cognitive scientists do is to study as precisely as possible each of the models being used, so that one can see, on any given occasion, the kind of reasoning being employed. That is one of the goals of this book. So far as I have been able to tell, there are two major categories of moral and political models for reasoning about politics— conservative and liberal. Most voters have some version of each model—and commonly apply different models to different issues at different times. The question asked here is: What are the models being applied?

One way of understanding recent electoral history is that voters have applied one model in presidential elections and the other in congressional elections, reproducing the Strict Father and Nurturant Mother in the national family. During the cold war, we had Strict Father presidents and Nurturant Mother Congresses. With the end of the cold war and the turn to domestic issues, the country first chose a Nurturant Parent president and then turned to a Strict Father Congress. Again, voters may not be consistent in the use of their models.

A strict conservative and a strict liberal each has coherent political views; they don't go back and forth between models from time to time and issue to issue. The models described here are those of strict conservatives and strict liberals. In short, the models define coherent ideologies. What conservative and liberal political leaders and ideologues do is to try to get voters to become coherent in their views—to move to one pole or the other, that is, to be entirely liberal or entirely conservative over the full range of issues.

Because people do not use the same models in all aspects of their lives, a political conservative could very well use the Nurturant Parent model in his family life but not in his political life, just as a political liberal could use Strict Father morality in his family life but the Nurturant Parent model in his political life. Strict fathers can be political liberals and nurturant parents can be political conservatives.

Contemporary conservative politics tries to link the family use and the political use of the models more closely; to point out that conservatives have the Strict Father model of the family and to convince others with the Strict Father model of the family that they should be political conservatives. I suspect that they are being successful at convincing people who believe in and identify with the Strict Father model of the family to vote conservative. For example, blue-collar workers who may previously have voted with liberals because of their union affiliation or economic interests may now, for cultural reasons, identify with conservatives and vote for them, even though it may not be in their economic interest to do so.

There is no logical contradiction between the use of one model as the basis of actual family life and the use of the other as the basis of one's politics. But logic is not what is at issue. There is more cognitive coherence and less cognitive dissonance if your politics is governed by the same model of the family that you use in your home life.

This book attempts to say what it means to be entirely liberal or entirely conservative. In working out the details, we will also see what it means to have a liberal or conservative view on each particular issue. In this way we will also be able to discern the forms of reasoning used on each issue by voters who go back and forth between models from time to time and issue to issue. As a result we can get a handle on the complexity of political thought.

WHAT IS EXPLAINED

One ought to ask a very pointed question at the outset: If you are going to characterize a great many variants of two models, as well as variations in the use of the models, what is explained? Given all that variation within categories and variation in the use of categories, can't you account for just about anything?

The very question reveals a misunderstanding of the enterprise. The point is not just to categorize. Classification in itself is relatively boring. The models do many things:

First, they analyze modes of reasoning.

Second, they show how modes of reasoning about different issues fit together.

Third, they show how different forms of, say, conservative reasoning are related to each other in such a way that they are all understood to be instances of conservatism.

Fourth, the models show the links between forms of political reasoning and forms of moral reasoning.

Fifth, the models show how moral reasoning in politics is ultimately based on models of the family.

And sixth, there must be an explanation of why the models fit together as they do—and therefore, why we don't merely have random forms of political reasoning. This is a tall order.

Cognitive science is, in itself, apolitical. Those of us who study conceptual systems do so simply to find out what we can about the workings of the human mind. But the same mind that we study for scientific reasons creates moral and political systems of thought and uses them every day. For this reason, the findings of conceptual-systems research will eventually come to matter more and more in understanding moral and political life. I see this book as an early step in the development of a cognitive social science that can allow us to comprehend our social and political lives better.

Personal Commitments

It is vital to be as clear as possible about the line between what one can discover about morality and politics using the tools of one's profession and what one's own moral and political commitments are. There are those who believe that drawing such a line is impossible, and maybe it is. But I am going to do my best anyway. In the first nineteen chapters of this book, I will be functioning as a cognitive scientist. I will be providing, to the best of my abilities, a cognitive analysis of the moral and political worldviews of conservatives and liberals in contemporary America, an analysis that I hope will be independent of any political prejudice.

But I cannot hide my own moral and political views and I will not try to. In the last few chapters of the book, I will give some reasons for why I am a liberal, reasons based not on liberal ideology itself but on external considerations.

I consider this book to be anything but an idle academic exercise. Because conservatives understand the moral dimension of our politics better than liberals do, they have been able not only to gain political victories but to use politics in the service of a much larger moral and cultural agenda for America, an agenda that if carried out would, I believe, destroy much of the moral progress made in the twentieth century. Liberals have been helpless to stop them, largely, I think, because they don't understand the conservative worldview and the role of moral idealism and the family within it.

Moreover, liberals do not fully comprehend the moral unity of their own politics and the role that the family plays in it. Liberals need to understand that there is an overall, coherent liberal politics which is based on a coherent, well-grounded, and powerful liberal morality. If liberals do not concern themselves very seriously and very quickly with the unity of their own philosophy and with morality and the family, they will not merely continue to lose elections but will as well bear responsibility for the success of conservatives in turning back the clock on progress in America. Conservatives know that politics is not just about policy and interest groups and issue-by-issue debate. They have learned that politics is about family and morality, about myth and metaphor and emotional identification. They have, over twenty-five years, managed to forge conceptual links in the voters' minds between morality and public policy. They have done this by carefully working out their values, comprehending their myths, and designing a language to fit those values and myths so that they can evoke them with powerful slogans, repeated over and over again, that reinforce those family-morality-policy links, until the connections have come to seem natural to many Americans, including many in the media. As long as liberals ignore the moral, mythic, and emotional dimension of politics, as long as they stick to policy and interest groups and issue-by-issue debate, they will have no hope of understanding the nature of the political transformation that has overtaken this country and they will have no hope of changing it.

THE TERM "LIBERAL"

There are many meanings for "liberal," some of which overlap with the meanings of "conservative." For the sake of clarity, I distinguish between "political liberalism," which is the subject of this book, and "theoretical liberalism," which is a topic in the field of political philosophy. I define "classical theoretical liberalism" as the view, which has a long history, that individuals are, or should be, free, autonomous rational actors, each pursuing their own self-interest.

On this account, many conservatives and libertarians are classical theoretical liberals.

Modern theoretical liberalism, on the other hand, arises primarily from the work of philosopher John Rawls (see References, sec. C3). Rawls sought to modify classic liberalism to include social issues, such as poverty, health, and education. He proposed the following social-contract theory of a just society (presented here in a much oversimplified fashion) to be added on to the classical view of the autonomous rational actor:

1. The Veil of Ignorance: The social contract must be drawn up as if no one knew where they were going to fit into society.
2. The result is that justice is seen as fairness. After all, if you don't know where you are going to fit into a society, you will want that society to be fair. If you were to wind up as low man on the totem pole, you would want that not to be so bad a position to be in.
3. An individual's choices of ends, values, and conceptions of the good are subjective expressions of preference. This makes them literal, rankable, and subject to mathematical theories of preference, utility, decision-making, etc.
4. Accepting this political view does not commit one to any particular moral view.
5. This view is universal and independent of particular cultures and subcultures.

Rawls's views have been elaborated and have been criticized on many grounds, especially on ''communitarian'' grounds, namely, that people are not just isolated autonomous individuals but (1) live in communities in which they have responsibilities, (2) are partly defined by those communities, (3) are partly defined by the ends they have and by conceptions of what morality is, and that (4) morality is a social phenomenon and meanings are social, not individual (see References, C4). All of this discussion is theoretical, rather than empirical, in nature. It is an attempt to characterize what liberalism should be, rather than what contemporary political liberalism is.

What I call "political liberalism," on the other hand, characterizes the cluster of political positions supported by people called "liberals" in our everyday political discourse: support for social programs; environmentalism; public education; equal rights for women, gays, and ethnic minorities; affirmative action; the pro-choice position on abortion; and so on. When I speak of "liberalism" in this book, I will be speaking of political liberalism, not theoretical liberalism.

But having made the distinction, I should ask the obvious question: Does modern theoretical liberalism in any form provide an accurate account of what political liberalism is? I think not, and the reasons should be clear by the end of the book. Political liberalism is a creature of a very different kind, and the worldview that characterizes it looks very different from any proposed form of theoretical liberalism.

Because this book studies empirical, not purely theoretical, issues, it should not be surprising that it comes up with a very different conception of "liberalism" than one finds in political philosophy. What is interesting is that what emerges has a few Rawls-like qualities and many qualities of the communitarian critiques. The rational actor shows up in the metaphor of Moral Self-Interest. Rawls's "veil of ignorance" has functional similarities to the metaphor of morality as empathy, which gives rise to the metaphor of morality as fairness (Chapter 6). Many of the communitarian views of liberalism emerge from Nurturant Parent morality, which stresses social responsibility, both social and individual ends, morality as being fundamentally social, and politics as fundamentally moral; other aspects of Nurturant Parent morality stress individual rights and freedoms.

I am not a political philosopher and did not begin with

any philosophical presuppositions about what I would find. Nor did I use either the intellectual tools or forms of reasoning of political philosophers. These results emerged from empirical study using the tools of a cognitive scientist to study political worldviews. As empirical findings, they have a very different status than theoretical speculations, and so should not be confused with political philosophy—for which, incidentally, I have great respect.

An Outline of the Book

Chapter 2 raises questions to be answered, questions about conservative and liberal ideologies. It asks why liberals and conservatives have the clusters of positions that they have and what forms of reason shape their forms of discourse. It also brings up a collection of puzzles: questions that each side has about the other.

Part II describes the family-based moral systems on which conservatism and liberalism are based and shows the enormous role played by metaphor in those moral systems. Chapter 3 lays out the basis for all metaphors for morality. Within Part II, Chapter 4 describes our most basic metaphor for morality. Chapters 5 and 6 describe the Strict Father and Nurturant Parent models of the family and the family-based moral systems they give rise to. At this point the groundwork has been laid for the application of these family-based systems of morality to politics.

Part III provides the link between the moral analysis and the political analysis. Chapter 7 explains why such an analysis is needed and why previous analyses have failed. Chapter 8 characterizes the explanatory nature of the model. And Chapter 9 describes the moral categories induced by the two family-based moralities.

Part IV is where the political explanation is carried out. Chapters 10 through 16 discuss a broad range of issues, from social programs and crime to abortion, showing the logic of the liberal and conservative stands on these issues and how each stand derives ultimately from a version of one of the two family-based moral systems.

Part V sums up the explanation. Chapter 17 surveys variations within the ranks of conservatives and liberals. Chapter 18 describes varieties of conservatism and liberalism that are "pathological" relative to the central models of the categories and discusses the issue of stereotyping. Chapter 19 argues that there cannot be a politics in America without the kinds of family-based moral systems described in the previous chapters.

Up until this point, the book provides a neutral description of conservative and liberal conceptual systems. In Part VI, I ask if there are any reasons *not* grounded in one of the ideologies for choosing between these moral and political systems. Chapters 21 through 23 provide three such reasons—reasons for being a liberal. Many other reasons might be adduced—and have been—but I bring up three that come from my field of research and adjacent fields: research on child development, the nature of mind, and the internal structure of moral conceptual systems. Finally, in the Epilogue, I discuss problems for public discourse and the media that are revealed by this study.

Overall, the book has a linear structure: First, the questions to be answered. Second, the first step in answering them, namely, the family-based moral systems. Third, the link between the moral systems and politics. Fourth, the politics and the answers to the questions. Fifth, nonideological reasons for choosing between the political worldviews. Sixth, implications for public discourse.

The Worldview Problem
for American Politics

Puzzles for Liberals

Conservatives are fond of suggesting that liberals don't understand what they say, that they just don't get it. The conservatives are right. The ascendancy of conservative ideology in recent years and, in particular, the startling conservative victory in the 1994 congressional elections have left liberals mystified about a great many things. Here are some examples.

William Bennett, a major conservative politician and intellectual leader, has put a major part of his efforts into moral education. He has written *The Book of Virtues,* an 800-page collection of classical moral stories for children, which has been on the best-seller lists for more than eighty straight weeks. Why do conservatives think that virtue and morality should be identified with their *political* agenda and what view of morality do they profess?

Family values and fatherhood have recently become central to conservative politics. What are those family values, what is that conception of fatherhood, and what do they have to do with politics?

The conservative Speaker of the House of Representatives, embracing family values, suggested that the children of welfare mothers be taken away from the only families they have known and be placed in orphanages. This sounded like a contradiction of family values to liberals, but not to conservatives. Why?

Conservatives are largely against abortion, saying that they want to save the lives of unborn fetuses. The United States has an extremely high infant-mortality rate, largely due to the lack of adequate prenatal care for low-income mothers. Yet conservatives are not in favor of government programs providing such prenatal care and have voted to eliminate existing programs that have succeeded in lowering the infant mortality rate. Liberals find this illogical. It appears to liberals that "pro-life" conservatives *do* want to prevent the death of those fetuses whose mothers *do not* want them (through stopping abortion), but *do not* want to prevent the deaths of fetuses whose mothers *do* want them (through providing adequate prenatal care programs). Conservatives see no contradiction. Why?

Liberals also find it illogical that right-to-life advocates are mostly in favor of capital punishment. This seems natural to conservatives. Why?

Conservatives are opposed to welfare and to government funds for the needy but are in favor of government funds going to victims of floods, fires, and earthquakes who are in need. Why isn't this contradictory?

A liberal supporter of California's 1994 single-payer initiative was speaking to a conservative audience and decided to appeal to their financial self-interest. He pointed out that the savings in administrative costs would get them the same health benefits for less money while also paying for health care for the indigent. A woman responded, "It just sounds wrong to me. I would be paying for somebody else." Why did his appeal to her economic self-interest fail?

Conservatives are willing to increase the budgets for the military and for prisons on the grounds that they provide protection. But they want to eliminate regulatory agencies whose job is to protect the public, especially workers and consumers. Conservatives do not conceptualize regulation as a form of protection, only as a form of interference. Why?

Conservatives claim to favor states' rights over the power of the federal government. Yet their proposal for tort reform will invest the federal government with considerable powers previously held by the states, the power to determine what lawsuits can be brought for product liability and securities fraud, and hence the power to control product safety standards and ethical financial practices. Why is this shift of power from the states to the federal government not considered a violation of states' rights by conservatives?

In these cases, what is irrational, mysterious, or just plain evil or corrupt to liberals is natural, straightforward, and moral to conservatives. Yet, the answers to all these questions are obvious if you understand the conservative worldview, as we shall see below.

Puzzles for Conservatives

Of course, most conservatives have just as little understanding of liberals. To conservatives, liberal positions seem outrageously immoral or just plain foolish. Here are some corresponding questions that conservatives have about liberal positions.

Liberals support welfare and education proposals to aid children, yet they sanction the murder of children by supporting the practice of abortion. Isn't this contradictory?

How can liberals claim to favor the rights of children, when they champion the rights of criminals, such as convicted child molesters? How can liberals claim empathy for victims when they defend the rights of criminals?

How can liberals support federal funding for AIDS research and treatment, while promoting the spread of AIDS by sanctioning sexual behavior that leads to AIDS? In defending gay rights, liberals sanction homosexual sex; they sanction teenage sex by advocating the distribution of condoms in schools; they sanction drug abuse by promoting needle exchange programs for drug users. How can liberals say they want to stop the spread of AIDS while they sanction practices that lead to it?

How can liberals claim to be supporters of labor when they support environmental restrictions that limit development and eliminate jobs?

How can liberals claim to support the expansion of the economy when they favor government regulations that limit entrepreneurship and when they tax profitable investments?

How can liberals claim to help citizens achieve the American dream when they punish financial success through the progressive income tax?

How can liberals claim to be helping people in need when they support social welfare programs that make people dependent on the government and limit their initiative?

How can liberals claim to be for equality of opportunity, when they promote racial, ethnic, and sexual favoritism by supporting affirmative action?

To conservatives, liberals seem either immoral, perverse, misguided, irrational, or just plain dumb. Yet, from the perspective of the liberal worldview, what seems contradictory or immoral or stupid to conservatives seems to liberals to be natural, rational, and, above all, moral.

The Worldview Problem for Cognitive Science

These sets of puzzles present a challenge to anyone who is concerned about the structure of contemporary political thought. To the cognitive scientist, they are important data.

The job of the cognitive scientist in this instance is to characterize the largely unconscious liberal and conservative worldviews accurately enough so that an analyst can see just why the puzzles for liberals are not puzzles for conservatives, and conversely. Any cognitive scientist who seeks to describe the conservative and liberal worldviews is constrained by at least two adequacy conditions.

First, the worldviews must make the collections of political stands on each side into two natural categories. For example, the liberal worldview analysis must explain why environmentalism, feminism, support for social programs, and progressive taxation fit naturally together for liberals, while the conservative worldview analysis must explain why their opposites fit together naturally for conservatives.

Second, any adequate descriptions of these two worldviews must show why the puzzles for liberals are not puzzles for conservatives, and conversely. As we shall see, this is anything but an easy problem and there are to my knowledge no previous solutions to it.

But there is a third, far more demanding, adequacy condition on the characterization of conservative and liberal worldviews. Those worldviews must additionally explain the topic choice, word choice, and discourse forms of conservatives and liberals. In short, those worldviews must explain just how conservative forms of reasoning make sense to conservatives, and the same for liberals. Moreover, they must explain why liberals and conservatives choose different topics to discuss and use different words in their discourse to discuss them. Furthermore they must explain why sometimes the same words have very different meanings when used by liberals and conservatives. As Rush Limbaugh is fond of saying, "Words have meanings." But they don't always have the same meanings to liberals and to conservatives, and where their meanings differ, those differences should be accounted for by differences in worldview. Let us consider some examples of what must be explained.

THE LANGUAGE OF CONSERVATISM

Conservatives like to make fun of liberals, claiming that liberals just don't speak their language. Again, the conservatives are right. There is a language of conservatism, and it's not just words. The words are familiar enough, but not what they mean. For example, "big government" does not just refer to the size of government or the amount spent by it. One can see the misunderstanding when liberals try to reason with conservatives by pointing out that increasing the amount spent on the military and prisons increases "big government." Conservatives laugh. The liberals have just misused the term. I have heard a conservative talk of "freedom" and a liberal attempt a rebuttal by pointing out that denying a woman access to abortion limits her "freedom" to choose. Again, the liberal has used a word that has a different meaning in the conservative lexicon.

Words don't have meanings in isolation. Words are defined relative to a conceptual system. If liberals are to understand how conservatives use their words, they will have to understand the conservative conceptual system. When a conservative legislator says, in support of eliminating Aid to Families with Dependent Children (AFDC), "It's alright to have a soft heart, but you've gotta have a strong backbone," one must ask exactly what that sentence means in that context, why that sentence constitutes an argument against continuing AFDC, and what exactly the argument is. In Dan Quayle's acceptance speech to the Republican convention in 1992, he said, in a rhetorical question arguing against the graduated income tax, "Why should the best people be punished?'' To make sense of this, one must know why rich people are ''the best people'' and why the graduated income tax constitutes ''punishment.'' In other conservative discourse, progressive taxation is referred to as ''theft'' and ''taking people's money away from them.'' Conservatives do not see the progressive income tax as ''paying one's fair share'' or ''civic duty'' or even ''noblesse oblige.'' Is there anything besides greed that leads conservatives to one view of taxation over another?

Here are some words and phrases used over and over in conservative discourse: character, virtue, discipline, tough it out, get tough, tough love, strong, self-reliance, individual responsibility, backbone, standards, authority, heritage, competition, earn, hard work, enterprise, property rights, reward, freedom, intrusion, interference, meddling, punishment, human nature, traditional, common sense, dependency, self-indulgent, elite, quotas, breakdown, corrupt, decay, rot, degenerate, deviant, lifestyle.

Why do conservatives use this constellation of words and phrases in arguing for political policies and exactly how do they use them? Exactly what unifies this collection, what forms it into a single constellation? A solution to the worldview problem must answer all these questions and more. It must explain why conservatives choose to talk about the topics they do, why they choose the words they do, why those words mean what they do to them, and how their reasoning makes sense to them. Every conservative speech or book or article is a challenge to any would-be description of the conservative worldview.

The same, of course, is true of the liberal worldview. Liberals, in their speeches and writings, choose different topics, different words, and different modes of inference than conservatives. Liberals talk about: social forces, social responsibility, free expression, human rights, equal rights, concern, care, help, health, safety, nutrition, basic human dignity, oppression, diversity, deprivation, alienation, big corporations, corporate welfare, ecology, ecosystem, biodiversity, pollution, and so on. Conservatives tend not to dwell on these topics, or to use these words as part of their normal political discourse. A description of the liberal and conservative worldviews should explain why.

As I mentioned above, conservatism and liberalism are not monolithic. There will not be a single conservative or liberal worldview to fit all conservatives or all liberals. Conservatism and liberalism are radial categories. They have, I believe, central models and variations on those models. I take as my goal the description of the central models and the descriptions of the major variations on those central models.

The Goals

The principal goal of this book is to describe the conservative and liberal worldviews with enough detail and accuracy to meet all the adequacy conditions we have just discussed. What I have found is that conservatives have a deeper insight into their worldview than liberals have into theirs. Conservatives talk constantly about the centrality of morality and the family in their politics, while liberals did not talk about these things until conservatives started winning elections by doing so. My findings indicate that the family and morality are central to both worldviews. But where conservatives are relatively aware of how their politics relates to their views of family life and morality, liberals are less aware of the implicit view of morality and the family that organizes their own political beliefs. This lack of conscious awareness of their own political worldview has been devastating to the liberal cause.

Of course, any adequate theory of political worldview in these terms will have to account as precisely as possible for the relationship between views of the family and morality on one hand and public policy on the other. I shall focus initially on conservative public policy and return later to liberal public policy. I also want, as well as I can, to understand the relationship between morality and political ideology. For example, to answer such questions as why conservatives do not use the ideas of social forces and class in their explanations and why liberals do; or why conservatives tend to prefer nature over nurture, why they tend to like books like *The Bell Curve,* while liberals prefer nurture over nature in their explanations.

In addition, I have found our public discourse about the nature of morality and its relation to politics to be sadly impoverished. We *must* find a way to talk about alternative moral systems and how they give rise to alternative forms of politics. Journalists—including the most intelligent and insightful of journalists—have been at a loss. They have to rely on existing forms of public discourse, and since those forms are not adequate to the task, even the most thoughtful and honest journalists need help. Public discourse has to be enriched so that the media can do its job better. I see this book as a step in the process of expanding our public discourse on the relationship between morality, politics, and family life. A major part of this effort is bringing to public discourse certain important ideas from the study of the mind. It is important that the public become aware that we think by using conceptual systems that are not immediately accessible to consciousness and that conceptual metaphor is part of our normal thought processes.

The Basic Claim

To date, I have found only one pair of models for conservative and liberal worldviews that meets all three adequacy conditions, a pair that (1) explains why certain stands on issues go together (e.g., gun control goes with social pro-

grams goes with pro-choice goes with environmentalism);
(2) explains why the puzzles for liberals are not puzzles for
conservatives, and conversely; and (3) explains topic choice,
word choice, and forms of reasoning in conservative and
liberal discourse. Those worldviews center on two opposing
models of the family.

At the center of the conservative worldview is a Strict
Father model.

This model posits a traditional nuclear family, with
the father having primary responsibility for supporting
and protecting the family as well as the authority to
set overall policy, to set strict rules for the behavior
of children, and to enforce the rules. The mother has
the day-to-day responsibility for the care of the house,
raising the children, and upholding the father's authority. Children must respect and obey their parents; by
doing so they build character, that is, self-discipline
and self-reliance. Love and nurturance are, of course,
a vital part of family life but can never outweigh parental authority, which is itself an expression of love
and nurturance—tough love. Self-discipline, self-
reliance, and respect for legitimate authority are the
crucial things that children must learn.

Once children are mature, they are on their own
and must depend on their acquired self-discipline to
survive. Their self-reliance gives them authority over
their own destinies, and parents are not to meddle in
their lives.

The liberal worldview centers on a very different ideal of
family life, the Nurturant Parent model:

Love, empathy, and nurturance are primary, and
children become responsible, self-disciplined and self-
reliant through being cared for, respected, and caring

for others, both in their family and in their community. Support and protection are part of nurturance, and they require strength and courage on the part of parents. The obedience of children comes out of their love and respect for their parents and their community, not out of the fear of punishment. Good communication is crucial. If their authority is to be legitimate, parents must explain why their decisions serve the cause of protection and nurturance. Questioning by children is seen as positive, since children need to learn why their parents do what they do and since children often have good ideas that should be taken seriously. Ultimately, of course, responsible parents have to make the decisions, and that must be clear.

The principal goal of nurturance is for children to be fulfilled and happy in their lives. A fulfilling life is assumed to be, in significant part, a nurturant life—one committed to family and community responsibility. What children need to learn most is empathy for others, the capacity for nurturance, and the maintenance of social ties, which cannot be done without the strength, respect, self-discipline, and self-reliance that comes through being cared for. Raising a child to be fulfilled also requires helping that child develop his or her potential for achievement and enjoyment. That requires respecting the child's own values and allowing the child to explore the range of ideas and options that the world offers.

When children are respected, nurtured, and communicated with from birth, they gradually enter into a lifetime relationship of mutual respect, communication, and caring with their parents.

Each model of the family induces a set of moral priorities. As we shall see below, these systems use the same moral principles but give them opposing priorities. The resulting moral systems, put together out of the same elements, but in different order, are radically opposed.

Strict Father morality assigns highest priorities to such things as moral strength (the self-control and self-discipline to stand up to external and internal evils), respect for and obedience to authority, the setting and following of strict guidelines and behavioral norms, and so on. Moral self-interest says that if everyone is free to pursue their self-interest, the overall self-interests of all will be maximized. In conservatism, the pursuit of self-interest is seen as a way of using self-discipline to achieve self-reliance.

Nurturant Parent morality has a different set of priorities. Moral nurturance requires empathy for others and the helping of those who need help. To help others, one must take care of oneself and nurture social ties. And one must be happy and fulfilled in oneself, or one will have little empathy for others. The moral pursuit of self-interest only makes sense within these priorities.

The moral principles that have priority in each model appear in the other model, but with lesser priorities. Those lesser priorities drastically change the effect of those principles. For example, moral strength appears in the nurturance model, but it functions not for its own sake, but rather in the service of nurturance. Moral authority, in the nurturance model, functions as a consequence of nurturance. Moral guidelines are defined by empathy and nurturance. Similarly, in the Strict Father model, empathy and nurturance are present and important, but they never override authority and moral strength. Indeed, authority and strength are seen as expressions of nurturance.

What we have here are two different forms of family-based morality. What links them to politics is a common understanding of the nation as a family, with the government as parent. Thus, it is natural for liberals to see it as the function of the government to help people in need and hence to support social programs, while it is equally natural for conservatives to see the function of the government as requiring citizens to be self-disciplined and self-reliant and, therefore, to help themselves.

This is just a mere hint of the analysis of the conservative and liberal worldviews. The details of the family models and the moral systems are far more complex and subtle and, correspondingly, so are the details of the political analysis. This overview is also too brief to discuss variations on the conservative and liberal positions. The full-blown analysis requires a lot more, beginning with a detailed account of our moral conceptual system.

HIDDEN VERSUS OVERT, DESCRIPTIVE VERSUS PRESCRIPTIVE

Before proceeding, it is crucial to put aside two common misunderstandings. The first is that many people believe that they are consciously aware of their own worldviews and that all one has to do to find out about people's views of the world is to ask them. Perhaps the most fundamental result of cognitive science is that this is not true. What people will tell you about their worldview does not necessarily accurately reflect how they reason, how they categorize, how they speak, and how they act. For this reason, someone studying political worldviews must establish adequacy conditions for an analysis, just as we have done. As we shall see, the kinds of things that conservatives and liberals say about their political worldviews do not meet these conditions of adequacy. If you ask a liberal about his political worldview, he will almost certainly talk about liberty and equality, rather than about a nurturant parent model of the family. But as we will see, such directly political ideas do not meet our adequacy conditions; they do not explain why the various liberal stands fit together, nor do they answer the puzzles or account for topic choice, language choice, and modes of reason. Where just asking people fails, as it usually does, the cognitive scientist turns to model-building, as I have in this book. The idea is to construct a model of unconscious political worldviews that will meet those adequacy conditions as closely as possible.

A second common misconception confuses description with prescription. The models we are discussing are *de*scriptive, not *pre*scriptive. They are attempts to describe what people's actual unconscious worldviews are, not what they should be. Most theories of liberalism and conservatism are not concerned with description but with prescription. For example, John Rawls's celebrated theory of liberalism is not an empirical descriptive study but an attempt to characterize a prescriptive theory of justice, from which liberalism follows. As a descriptive account of actual liberal political stands on issues, it is a failure, as we shall see. My job here is to describe how people do make sense of their politics, not how they should.

The same goes for the account of morality that I am about to give. I am interested not in what morality should be, but in how our very notions of what is moral are built into our unconscious conceptual systems.

Part Two

Moral Conceptual Systems

3

Experiential Morality

To understand the ways in which moral worldviews affect political worldviews, we must look first at our system of moral concepts in some detail. Because I will argue that political perspectives are derived from systems of moral concepts, we must consider what those concepts typically consist of and why we have the moral concepts we have.

An important conclusion of research in cognitive studies is that moral thinking is imaginative and that it depends fundamentally on metaphorical understanding (see References, A6; Johnson 1993). Before we proceed with our discussion of metaphors for morality, we should point out the obvious—that morality is not all metaphorical and that nonmetaphorical aspects of morality are what the system of metaphors for morality is based on. Nonmetaphorical morality is about the experience of well-being. The most fundamental form of morality concerns promoting the experiential well-being of others and the avoidance and prevention of experiential harm to others or the disruption of the well-being of others.

Here is part of what is meant by "well-being": Other things being equal, you are better off if you are *healthy* rather than sick, *rich* rather than poor, *strong* rather than weak, *free* rather than imprisoned, *cared for* rather than uncared for, *happy* rather than sad, *whole* rather than lacking, *clean* rather than filthy, *beautiful* rather than ugly, if you are functioning in the *light* rather than the dark, if you can stand *upright* so that you don't fall down, and if you live in a community with *close social ties,* rather than in a hostile or isolated one. These are among our basic experiential forms of well-being. Their opposites are forms of harm or lack of well-being: poverty, illness, sadness, weakness, imprisonment, and so on. Immoral action is action that causes harm or lack of well-being, that is, action that deprives someone of one or more of these—of health, wealth, happiness, strength, freedom, safety, beauty, and so on. In the case of young children, it is the job of parents to do their best to guarantee their well-being. On the whole, young children are better off if they are *obedient* rather than disobedient to their parents, who, in the normal case, have their best interests at heart, know how to keep them from being harmed, and exercise legitimate authority.

These are, of course, norms, and the qualification ''other things being equal'' is necessary, since one can think of special cases where these may not be true. A wealthy child may not get the necessary attention of its parents, someone beautiful may be the target of envy, one ordinarily needs to be in the dark in order to sleep, excessive freedom can sometimes be harmful, sadness and pain may be necessary to appreciate happiness, social ties that are too close can become oppressive, and parents may be abusive or neglectful or ignorant. But, on the whole, this account of experiential well-being holds.

It is these conditions that form the grounding for our system of moral metaphors. Since it is better to be rich than to be poor, morality is conceptualized in terms of wealth. Since it is better to be strong than to be weak, we expect to see morality conceptualized as strength. Because it is better to be healthy than sick, it is no surprise to see morality conceptualized in terms of health and attendant concepts like cleanliness and purity. Since it is better to be cared for than uncared for, it seems natural to find morality conceptualized as nurturance. And because, in normal cases, children tend to be better off if they obey rather than disobey their parents, we expect to see morality conceptualized as obedience.

What we learn from this is that metaphorical morality is grounded in nonmetaphorical morality, that is, in forms of well-being, and that the system of metaphors for morality as a whole is thus far from arbitrary. Because the same forms of well-being are widespread around the world, we expect many of the same metaphors for morality to show up in culture after culture—and they do. Where we find purification rituals, we find a manifestation of Morality as Purity. Because of the widespread fear of the dark, we find a widespread conception of Evil as Dark and Good as Light. Because it is better to walk upright than to fall down, we find the widespread metaphor of Morality as Uprightness. In short, because our notion of what constitutes well-being is widely shared, our pool of metaphors for morality is also widely shared. Indeed, the commonality of shared metaphors for morality both within and across societies raises a deep question: What are the differences in moral systems and what is the source of those differences? As we shall see, different conceptions of the family constitute one such source of difference—at least within American culture.

4

Keeping the Moral Books

The metaphors we use in discussing moral questions are abundant. We use these metaphors to frame moral issues: to interpret them, understand them, and explore their consequences. We will see that they play an absolutely central role in our judgments about what is good behavior and what is bad, what is the right thing to do and what is wrong.

There is at least one class of metaphors for morality, however, that do not in themselves tell you exactly which actions are moral or immoral. They are in that sense *metamoral*. When combined with other metaphors, they generate moral conclusions about various kinds of behavior.

One of the most important metamoral concepts concerns how moral books are kept. It is this metaphor through which we understand concepts like justice, fairness, retribution, and revenge.

The Moral Accounting Metaphor

We all conceptualize well-being as wealth. We understand an increase in well-being as a "gain" and a decrease of well-being as a "loss" or a "cost." When we speak of the "costs" of a fire or an earthquake, we do not mean just the monetary cost but also the "cost" in human well-being— deaths, injuries, suffering, trauma. When we speak of "profiting" from an experience, we are speaking of the kinds of well-being we might "gain" from that experience— perhaps knowledge, enjoyment, sophistication, or confidence.

The metaphor by which we conceptualize well-being as wealth is a metaphor that is ubiquitous and important in our conceptual systems. Whenever we are not talking literally about money, and we ask whether a course of action is "worth it," we are using this financial metaphor to treat the resulting well-being or harm as if they were money and to see if the course of action is sufficiently "profitable." This economic metaphor allows us to use the ways that we think and talk about money to think and talk about well-being. Most important, it allows us to think about something qualitative (well-being) in terms of something quantitative (money), which in turn allows us to bring to bear our extensive forms of quantitative reasoning on something as elusive as well-being.

The metaphor of Well-Being As Wealth can be combined with a very general metaphor for causal action in which causation is seen as object transfer—as the giving of an effect to an affected party, as in "The noise *gave* me a headache." The conception Well-Being As Wealth allows us to see an effect that helps as a gain and one that harms as a loss. Under normal circumstances, moral action is seen as action intended to help (provide a gain) and immoral action as action intended to harm (provide a loss). By this conceptual mechanism, an action of moral import is conceptualized in terms of a financial transaction, with a moral interaction being metaphorically equivalent to a financial transaction, one in which the books are balanced. Just as literal bookkeeping is vital to economic functioning, so moral bookkeeping is vital to social functioning. And just as it is important that the financial books be balanced, so it is important that the moral books be balanced.

Of course, the source domain of the metaphor, the domain of financial transaction, itself has a morality: It is moral to pay your debts and immoral not to. When moral action is understood metaphorically in terms of financial transaction, financial morality is carried over to morality in general: There is a moral imperative not only to pay one's financial debts but also one's moral debts.

The Moral Accounting Schemes

The general metaphor of Moral Accounting is realized in a small number of basic moral schemes: reciprocation, retribution, restitution, revenge, altruism, and others. Each of these moral schemes is defined using the metaphor of Moral Accounting, but the schemes differ to how they use this metaphor, that is, they differ as to their inherent logics. Here are the basic schemes.

RECIPROCATION

If you do something good for me, then I "owe" you something, I am "in your debt." If I do something equally good for you, then I have "repaid" you and we are even. The books are balanced.

From the perspective of cognitive science, there is much here to be explained. Why are financial words like "owe," "debt," and "repay" used to speak of morality? And why is the logic of gain and loss, debt and repayment used to *think* about morality?

The answer given here is that we use conceptual metaphors to think about aspects of morality and that among the general conceptual metaphors used to think and talk about morality are Well-Being As Wealth, and Moral Action As Financial Transaction. But the discovery of these conceptual metaphors (see References, A1, Taub 1990; Klingebiel 1990; A6, Johnson 1993) raises still another question. Why should well-being be conceptualized as wealth? The answer was discussed in the previous chapter. Our metaphors for morality rest on our notions of basic experiential well-being. Wealth forms one basis for a metaphor for morality and, as we shall see, health, strength, and other forms of experiential well-being lead to other metaphors for morality.

But let us return to the moral scheme of reciprocation, where your doing something good for me places me in your "debt," and I can repay what I "owe" by doing something equally good for you.

Even in this simple case, there are two principles of moral action. *The positive-action principle:* Moral action is giving something of positive value; immoral action is giving something of negative value. *The debt-payment principle:* There is a moral imperative to pay one's moral debts; the failure to pay one's moral debts is immoral. Thus, when you did something good for me, you engaged in positive action, which is moral. When I did something equally good for you, I engaged in *both* forms of moral action. I did something good for you *and* I paid my debts. Here the two principles act in concert.

RETRIBUTION

Suppose someone does something to harm you and you say, "I'll pay you back—with interest!" What you are doing is threatening retribution. Why, exactly, should a statement about repayment with interest be a threat of retribution? To see why, we have to look at moral transactions where the values are negative.

Moral transactions get complicated in the case of negative

action. The complications arise because moral accounting is governed by a moral version of the arithmetic of keeping accounts, in which gaining a credit is equivalent to losing a debit and gaining a debit is equivalent to losing a credit.

Suppose I do something to harm you. Then, by Well-Being As Wealth, I have given you something of negative value. You owe me something of equal (negative) value. By moral arithmetic, giving something negative is equivalent to taking something positive. By harming you, I have taken something of value from you. Are you going to "let me get away with it"?

By harming you, I have placed you in a potential moral dilemma with respect to the first and second principles of moral accounting. Here are the horns of the dilemma:

The first: If you now do something equally harmful to me, you have done something with two moral interpretations. By the positive-action principle, you have acted immorally since you did something harmful to me. ("Two wrongs don't make a right.") By the debt-payment principle, you have acted morally, since you have paid your moral debts.

The second: Had you done nothing to punish me for harming you, you would have acted morally by the positive-action principle, since you would have avoided doing harm. But you would have acted immorally by the second principle: in "letting me get away with it" you would not have done your moral duty, which is to "make me pay" for what I have done. You would not have lived up to the debt-payment principle.

No matter what you do, you violate one of the two principles. You have to make a choice. You have to give priority to one of the principles. Such a choice gives two different versions of moral accounting: The Morality of Absolute Goodness puts the first principle first. The Morality of Retribution puts the second principle first. As might be expected, different people and different subcultures have different solutions to this dilemma, some preferring retribution, others preferring absolute goodness.

In debates over the death penalty, liberals rank Absolute Goodness over Retribution, and conservatives tend to prefer Retribution: a life for a life.

Suppose again that you do something to harm me, which is metaphorically to give me something of negative value. Moral arithmetic provides another version of retribution. By moral arithmetic, you have taken something of positive value from me by harming me. If I take something of equal positive value back from you, that too is retribution, another way of balancing the moral books.

Up to now, I have used the word "retribution" in too loose a fashion. Actually, "retribution" is limited to cases where the balancing of the moral books is carried out by some legitimate authority. "Revenge" occurs when the same principle is used without legitimate authority.

Thus, when a father spanks a child for being naughty, it is retribution, not revenge. When a court sentences a convicted criminal to jail, it is retribution, not revenge. But when a man takes the law into his own hands and shoots his brother's murderer, that is revenge.

RESTITUTION

If I do something harmful to you, then I have given something of negative value and, by moral arithmetic, taken something of positive value. I then owe you something of equal positive value. I can therefore make restitution—make up for what I have done—by paying you back with something of equal positive value. Of course, in many cases, full restitution is impossible, but partial restitution may be possible.

An interesting advantage of restitution is that it does not place you in a moral dilemma with respect to the positive-action and debt-payment principles. You both perform a positive action and you pay your debt.

ALTRUISM

If I do something good for you, then by moral accounting I have given you something of positive value. You are then in my debt. In altruism, I cancel the debt, since I don't want anything in return. I nonetheless build up moral "credit."

The concept of moral credit arises from moral accounting. In a moral system, the moral books must be balanced. Therefore, if you have a credit to your account in the form of a debt from someone else, the credit doesn't just disappear when you cancel the debt. Instead, it becomes a moral credit. For someone to be a good person, he has to have a lot of moral credit. A great deal of morality has to do with the principles by which you amass moral credit.

TURNING THE OTHER CHEEK

If I harm you, I have (by Well-Being As Wealth) given you something of negative value, and (by moral arithmetic) taken something of positive value. Therefore, I owe you something of positive value. Suppose you then refuse both retribution and revenge. You either allow me to harm you further or, perhaps, you even do something good for me. By moral accounting, either harming you further or accepting something good from you would incur even further debt: by turning the other cheek, you make me even more morally indebted to you. If I have a conscience, then I should feel even more guilty. Turning the other cheek involves the rejection of retribution and revenge and the acceptance of basic goodness—and when it works, it works via this mechanism of moral accounting.

KARMA: MORAL ACCOUNTING WITH THE UNIVERSE

The Buddhist theory of Karma has a contemporary American counterpart: What goes around comes around. The basic idea is that you can affect the balance of good and bad things that happen to you by your actions: You will get what you deserve. The more good things you do for people, the more good things will happen to you. The more bad things you do to people, the more bad things will happen to you.

In another version of moral balance with the universe, the good and bad things that happen to you are balanced out. Thus, we occasionally find people saying things like, "Things have been rotten for a long time. They're bound to get better," or "Too many good things have been happening to me. I'm starting to get scared."

REWARD AND PUNISHMENT

Another basic schema using the metaphor of moral acounting is that of reward and punishment. The basic reward-punishment schema is one in which one person has authority over another. A reward is reciprocation by a person in authority, while a punishment is retribution by a person in authority. As in reciprocation and retribution, performing an act for someone's benefit is conceptualized metaphorically as giving something of positive value, and performing an act to someone's detriment is seen as giving something of negative value (or taking something of positive value).

Thus, a father may reward his child for helping him clean out the garage, but a child doesn't reward his father for helping him clean out his room. Similarly, a father may punish a child for throwing plums at passing cars, but if a child hits his father with a toy for refusing to let him stay up late, the child isn't punishing the father (unless the father has ceded him authority).

Rewards and punishments are moral acts; giving someone an appropriate reward or punishment balances the moral books. An important special case arises when the person in authority gives an order. That order imposes an obligation to obey. The obligation to obey is a metaphorical debt. You *owe* obedience to someone who has authority over you. If you obey, you are paying the debt; if you don't obey, you are refusing to pay the debt—an immoral act, equivalent by moral arithmetic to stealing, a crime. When you disobey a legitimate authority, it is moral for you to be punished, to receive something of negative value or have something of positive value taken from you. Moral accounting then, of course, says that the punishment must fit the crime.

But actually disobedience involves *two* crimes. To see why, recall that, in reciprocation, there are two moral acts, determined by two principles: *The positive-action principle: Moral action is giving something of positive value; immoral action is giving something of negative value. The debt-payment principle:* There is a moral imperative to pay one's moral debts; the failure to pay one's moral debts is immoral. Disobedience violates both principles. It is both a refusal to engage in positive action and a refusal to pay your debts. The first violation is specific: you were obligated to do something specific and you didn't do it. But the second violation is general: it is a violation of the whole system of authority in terms of which obedience is defined.

In a system where authority is of the utmost importance, the debt-payment principle is given much heavier weight than the positive-action principle. Consider an example. Suppose you are a private in the army and your sergeant throws a shoe on the floor and orders you to pick it up. If you refuse to pick up the shoe, you are violating two obligations—the obligation to pick up the shoe and the obligation to obey orders. Not picking up the shoe is a minor thing in itself, but not obeying orders is a major offense—an offense against the entire authority-based system of the army. In the army, it is moral for the sergeant to punish the private heavily for challenging the basis of the entire system. Indeed, some sergeants may see such punishment as a moral duty, a duty to uphold the most fundamental principle that allows the system to operate.

In a common interpretation of the Old Testament, God punished Adam and Eve for taking bites out of a fruit from the tree of the knowledge of good and evil. The punishment was banishment from the Garden of Eden and the loss of immortality for them and their descendants; that is, the punishment was death and suffering for all mankind forever. Now that might seem a little extreme for taking two bites out of a piece of fruit. But the issue wasn't the fruit itself; the issue was disobedience—the violation of the debt-payment principle upon which all authority rests. Eating the fruit was not just eating a fruit, it was metaphorically a loss of innocence and a giving in to the temptation to satisfy one's desires—giving in to the temptations of the flesh. It showed that all people are susceptible to temptations of the flesh, to a loss of innocence, and therefore to disobedience, which challenges the entire principle of God's legitimate authority.

God relented somewhat and offered back immortality and freedom from suffering, not to all mankind, but on a one-by-one basis. Each person who accepted God's legitimate authority and who became sufficiently self-disciplined to overcome the temptations of the flesh and obey God's commandments for the remainder of his life on earth would get everlasting life and relief from suffering in heaven. That would be God's reward for obedience. A person who could do that would have earned that reward. In short, one common interpretation of the Judeo-Christian tradition rests on the metaphor of Moral Accounting, on the ideas of reward and punishment, and on the two principles of positive action and debt-payment. An interpretation that places authority above all else will give much greater weight to the debt-payment principle than to the positive-action principle.

There are, of course, other interpretations in which God's grace is given heavier weight than his authority, that is, where positive action is given heavier weight than obedience, which is a form of debt-payment (see chapter 14).

WORK

There are two different common metaphors for work, each of which uses moral accounting. We will call them the Work Exchange metaphor and the Work Reward metaphor. In the Work Reward metaphor, the employer is conceptualized as having legitimate authority over the employee, and pay is a reward for work. The metaphor can be stated as follows:

- The employer is a legitimate authority.
- The employee is subject to that authority.
- Work is obedience to the employer's commands.
- Pay is the reward the employee receives for obedience to the employer.

This metaphor makes work a part of the moral order—a hierarchical chain of legitimate authority. This conception of work implies the following:

- The employer has a right to give orders to the employee, and to punish the employee for not obeying those orders.
- Obedience is the condition of employment.
- The social relationship of employer to employee is one of superior to inferior.
- The employer knows best.
- The employee is moral if he obeys the employer.
- The employer is moral if he appropriately rewards the employee for obeying his orders.

In the Work Exchange metaphor, work is seen as an object of value. The worker voluntarily exchanges his work for money. The metaphor can be stated as follows:

- Work is an object of value.
- The worker is the possessor of his work.
- The employer is the possessor of his money.
- Employment is the voluntary exchange of the worker's work for the employer's money.

In the context of labor unions and contracts, the nature and value of the work are mutually agreed on in the contract. Payment is a matter of agreed upon exchange, not reward. Work is a matter of trade, not obedience. The nature and limits of authority are spelled out in the contract.

Both of these conceptualizations of work depend upon the metaphor of Moral Accounting—in the first case to define appropriate reward, in the second case to determine the value of the work. Both conceptions are metaphorical, though they may seem literal if everyone involved agrees to abide by one metaphor or the other. What these metaphors show is that the concept of work is not absolute; it varies with the metaphors used to conceptualize it. They also show that work is part of a network of moral concepts, including moral accounting.

Some Basic Moral Concepts

The moral accounting metaphor allows us to see better how we conceptualize fundamental moral notions. As examples of concepts defined relative to moral accounting, we will consider trust, credit, justice, rights, duties, and self-righteousness.

CREDIT AND TRUST: MORAL CAPITAL

By moral accounting, a moral action is part of an exchange. When you do something good for someone, you give something of positive value to him and what you get in exchange is "credit." Credit for acting morally can accumulate. It is a form of capital.

What does it mean to trust someone with money? It means that you give him money with the confidence that he will give it back when you need it, or at an agreed upon time. To place your trust in someone morally is to give him advance moral credit, credit he has not yet earned, on the assumption that he will repay you by acting morally. If someone that you place your trust in acts very immorally, then you "lose trust" in him, that is, you lose the moral credit that you gave him in advance as prepayment for acting morally. He is "discredited" and "morally bankrupt." "Trust" is a prepayment of moral credit for future moral action. But in general, people do not trust just anyone. To be trusted, a person has to "build trust," to establish a history of being trustworthy, a moral credit rating.

JUSTICE

In the Moral Accounting metaphor, justice is the settling of accounts, which results in the balancing of the moral books. Justice is done when people get what they "deserve," when moral debits and credits cancel each other out.

RIGHTS AND DUTIES

The metaphorical conception of rights in Moral Accounting can be seen very clearly in Martin Luther King, Jr.'s classic "I Have A Dream" speech. The financial expressions concerning rights are in italics.

> In a sense we have come to our nation's capital to *cash a check*. When the architects of our republic

wrote the magnificent words of the Constitution and the Declaration of Independence, they were signing a *promissory note* to which every American was to *fall heir.* This note was a promise that all men—yes, black men as well as white men—would be guaranteed the inalienable rights of life, liberty, and the pursuit of happiness.

It is obvious today that America has *defaulted on this promissory note* insofar as her citizens of color are concerned. Instead of honoring this sacred obligation, America has given the Negro people a *bad check, a check which has come back marked "insufficient funds"* in the *great vaults* of opportunity of this nation. So we have come to *cash this check—a check that will give us on demand the riches* of freedom and the security of justice.

Rights, in the financial domain, are rights to one's property. If a bank is keeping your money, you have a right to get it. If someone has borrowed money from you, you have a right to get it back. The metaphor Well-Being As Wealth applies to this financial notion of a right to one's wealth, yielding a notion of a broader right, a right to one's well-being—special cases being life, liberty, and the pursuit of happiness. In short, a right to well-being in general is metaphorically understood as a right to wealth. That is why Dr. King could speak of the Constitution and Declaration of Independence, which guarantee rights, as *a promissory note.* If you have a right to opportunity, then coming to Washington to demand opportunity can be conceived of as coming to *cash a check.*

This understanding of rights is not mere rhetoric. It is one of two understandings that we have of rights in general. If rights in general are understood as rights to property, then there are two special cases: If the property is land, the right is a right of way, that is, a right of access to your property. If the property is money, it is the right to have your money given to you.

Let us see what follows from the metaphorical concept of a right as right to one's money. If that's what a right is, what is a duty? A duty is a metaphorical debt, something that is due to someone else, something that you have to pay. Duties can be of either a positive or negative nature, things you must do or things you must refrain from doing despite a possible desire or other reason to do so. Carrying out a duty, whether through action or refraining from action, is doing a form of moral work that counts as paying a moral debt. A failure to do one's duty is akin to a failure to pay a debt, which by moral arithmetic is a metaphorical form of stealing. Rights in general, like rights to property, can be either earned or inherited.

Rights and duties fit together: whenever someone has a right, someone else has a duty, and conversely. If you have a right to an education, someone has a duty to provide it. If you have a right to free speech, others have a duty to protect that speech or refrain from interfering with it. You cannot have a right to breathe clean air unless others have a duty to refrain from polluting it or to guarantee that there is clean air to breathe. In many cases, it falls to government to perform the duties that make rights possible. Where the duty of guaranteeing rights falls to the government, those rights are "purchased" through taxation. Lower taxes may mean fewer rights. If you want rights, somebody's got to pay for them or provide them. Rights and duties don't come into existence out of nothing. They require social, cultural, and political institutions and require at least metaphorical economics and often literal economics.

Because rights and duties are interdefined via moral accounting, it follows that the more rights people get, the more duties people get. The duties are often literal debts, as well as metaphorical debts—taxes that need to be paid to purchase those rights.

But it is not always clear that when certain people get more rights, the same people get more duties. This has led to what has been called the communitarian critique of rights. The critique is also based on moral accounting. It says (in the broadest terms) that having rights given to you by the community confers upon you duties to the community. If a right is a letter of moral credit and a duty is a moral debit, the credits and debits should balance out.

SELF-RIGHTEOUSNESS

A self-righteous person is someone who carefully keeps his own moral ledger books, who makes sure that, according to his own system of moral accounting, he is morally solvent and that, in his accounting system, his credits always outweigh his debits. A thoroughly self-righteous person knows neither shame nor gratitude, since he has no moral debts, again according to his own method of accounting.

There are three things that make him not righteous but *self-righteous*. The first is that he recognizes no moral values other than his own as valid. The second is that he keeps his own books. There is no external auditing. And the third is that he must communicate his moral standing to his interlocutors.

The self-righteous person's superfluity of moral credit is the basis of his discourse. He presupposes his own moral values and his own righteousness as a condition of conversation. The effect of this is that anyone talking to a self-righteous person must either agree with his moral values and act equally self-righteous, or face being put in a morally inferior position in the discourse. This is what makes self-righteous people particularly infuriating to talk to.

FAIRNESS

Children learn very early what is and isn't fair. Fairness is when the cookies are divided equally; when everybody gets a chance to play; when following the rules of the game allows for an equal chance at winning; when everybody does his job; and when you get what you earn or what you agree to. Unfairness is not getting as many cookies as your brother; not getting a chance to play; cheating, or bending the rules of the game to increase your chances of winning; not doing your job and therefore making others do it for you; or not getting what you earn or what is agreed upon.

In short, fairness is about the equitable distribution of objects of value (either positive or negative value) according to some accepted standard. What is distributed may be material objects—say cookies or money—or metaphorical objects, such as chances to participate, opportunities, tasks to be done, punishments or commendations, or the ability to state one's case.

There are many models of fairness:

- Equality of distribution (one child, one cookie)
- Equality of opportunity (one person, one raffle ticket)
- Procedural distribution (playing by the rules determines what you get)
- Rights-based fairness (you get what you have a right to)
- Need-based fairness (the more you need, the more you have a right to)
- Scalar distribution (the more you work, the more you get)
- Contractual distribution (you get what you agree to)
- Equal distribution of responsibility (we share the burden equally)

- Scalar distribution of responsibility (the greater your
  abilities, the greater your responsibilities)
- Equal distribution of power (one person, one vote)

Here procedural fairness is the impartial rule-based distribution of opportunities to participate, talk, state one's case, and so on.

One of the most basic conceptions of morality we have conceptualizes moral action as fair distribution and immoral action as unfair distribution. However, different versions of what constitutes fairness result in different versions of Morality As Fairness. Equality of distribution is very different from equality of opportunity. Rule-based fairness invites dispute over how impartial the rules really are. Rights-based fairness differs according to one's conception of what counts as a right. The communist slogan, "From each according to his abilities, to each according to his needs," is a composite of two schemas: the scalar distribution of responsibility and need-based distribution. It therefore invites *two* challenges: Is need-based distribution moral? And is the scalar distribution of responsibility moral? In short, seeing morality as fair distribution raises another set of thorny questions.

Another issue in fairness is what is to count as an instance of distribution in deciding upon an equitable distribution. Is it distribution over individuals or over groups defined by race, ethnicity, or gender? Is it a single act of distribution or multiple acts? Is it distribution at a single time or over a period of history? Disputes over whether affirmative action is fair (and hence moral) are disputes about such matters. Both sides in affirmative action assume the concept of Moral Action As Fair Distribution, but differ over such issues. In general, conservatives and liberals agree that Moral Action Is Fair Distribution, but they disagree strongly about what counts as fair distribution, for reasons that we shall see below.

Fairness necessarily involves a form of moral accounting, though it is the keeping of a different kind of moral books. What is distributed is taken to be something of positive value, something that may be actual wealth (in the case of taxation) or something that increases your well-being, and hence is treated as wealth by the Well-Being As Wealth metaphor. To guarantee fairness, somebody has to keep track of who gets what.

There is a language that goes with Moral Action As Fair Distribution. "That's not fair!" is a classic accusation that an immoral decision has been made. "You're not playing by the rules" or "You're cheating" are also accusations of unfair, and hence immoral, behavior. The same goes for "She's been denied her rights" or "He got something he didn't deserve." Any time these expressions are used, morality is being conceptualized as fair distribution, but the type of fair distribution differs from case to case.

Summary

There is a fundamental economic metaphor behind much of morality, the ubiquitous conception of Well-Being As Wealth, which brings quantitative reasoning into the qualitative realm of morality. It is so fundamental a metaphor that it is rarely even noticed as being metaphoric. Linguistically, the metaphor is made manifest by the use of economic words like *owe*, *debt*, and *pay* in the moral domain. Logically, the metaphor shows itself in the use in the moral domain of such quantitative forms of reason as moral arithmetic, which is taken from accounting.

There are two distinct general uses of the Well-Being As Wealth metaphor, one concerning the results of interaction and the other the results of distribution. The former is the

Moral Accounting metaphor, with special cases such as reciprocation, retribution, restitution, reward and punishment. The latter is Moral Action As Fair Distribution, with many versions of what is to count as fair distribution. Such moral schemes are not usually thought of as metaphorical, because the Well-Being As Wealth metaphor is so commonplace in everyday life. The metaphorical nature of these moral schemes is revealed in the transfer of words and forms of reason from the quantitative financial domain to the qualitative moral domain. These forms of metaphorical thought and language are both conventional and normal. The use of metaphorical thought and language in moral reasoning and discourse in no way impugns the metaphorical moral schemes involved. It does, however, serve to remind us that these are commonplace products of the human mind, not principles built into the objective structure of the universe.

This example should serve to introduce the notion of conceptual metaphor, as it has come to be used in contemporary cognitive linguistics. A conceptual metaphor is a correspondence between concepts across conceptual domains, allowing forms of reasoning and words from one domain (in this case, the economic domain) to be used in the other (in this case, the moral domain). It is extremely common for such metaphors to be fixed in our conceptual systems, and thousands of such metaphors contribute to our everyday modes of thought. For the most part, we use them without effort or conscious awareness. Yet, as we shall see, they play an enormous role in characterizing our worldviews.

In the next two chapters, we examine two family-based moral systems: strict father morality and nurturant parent morality. The point of each chapter is to present a collection of metaphors of morality that are prioritized differently by each of those models of the family. The metaphors have been arrived at independently by looking at two kinds of evidence: (1) how the language of a nonmoral domain (e.g., the financial domain) is used to talk about the moral domain, and (2) how the forms of reason used in nonmoral domains (e.g., health, strength) are used to reason about the moral domain.

The two models of the family are common stereotypes within our culture and should be immediately recognizable. The claims made in these chapters are these: The metaphors for morality that are cited do exist in the conceptual systems of Americans. The models of the family described are common models of what family life should be. Each model of the family motivates giving priority to the metaphors cited over other metaphors for morality. The result is two different family-based moral systems. The different metaphors of morality in each system give rise to different forms of moral reasoning.

At this point in the book nothing is claimed about the relevance of these family-based models of morality to political life. I will argue in later chapters that these pairings of family models with metaphors for morality are, in fact, central to the conservative and liberal conceptual worldviews. When we get to that discussion, the basis for those claims will be that they meet the adequacy conditions set forth in Chapter 2. But for now, we are merely outlining two organizations of moral priorities.

5

Strict Father Morality

In this chapter and the next, we will see that two different models of ideal family life can motivate corresponding sets of metaphorical priorities, each of which constitutes a distinct moral system. Let us begin with the following model of an ideal family, a model which Americans should find familiar. Different individuals may have somewhat different versions of it, but in its major outlines it is an important part of American mythology. Some of the variations on the model will be discussed at the end of the chapter.

The Strict Father Family

The Strict Father model takes as background the view that life is difficult and that the world is fundamentally dangerous. As Oliver North said repeatedly in his testimony to Congress, "The world is a dangerous place." Survival is a major concern and there are dangers and evils lurking everywhere, especially in the human soul. Here is the model:

A traditional nuclear family, with the father having primary responsibility for supporting and protecting the family as well as the authority to set overall family policy. He teaches children right from wrong by setting strict rules for their behavior and enforcing them through punishment. The punishment is typically mild to moderate, but sufficiently painful. It is commonly corporal punishment—say, with a belt or a stick. He also gains their cooperation by showing love and appreciation when they do follow the rules. But children must never be coddled, lest they become spoiled; a spoiled child will be dependent for life and will not learn proper morals.

The mother has day-to-day responsibility for the care of the house, raising the children, and upholding the father's authority. Children must respect and obey their parents, partly for their own safety and partly because by doing so they build character, that is, self-discipline and self-reliance. Love and nurturance are a vital part of family life, but they should never outweigh parental authority, which is itself an expression of love and nurturance—tough love. Self-discipline, self-reliance, and respect for legitimate authority are the crucial things that a child must learn. A mature adult becomes self-reliant through applying self-discipline in pursuing his self-interest. Only if a child learns self-discipline can he become self-reliant later in life. Survival is a matter of competition, and only through self-discipline can a child learn to compete successfully.

The mature children of the Strict Father have to sink or swim by themselves. They are on their own and have to prove their responsibility and self-reliance. They have attained, through discipline, authority over themselves. They have to, and are competent to, make their own decisions. They have to protect themselves and their families. They know what is good for them better than their parents, who are distant from them. Good parents do not meddle or interfere in their lives. Any parental meddling or interference is strongly resented.

I should say at the outset that, though I have used the term "Strict Father" to name the model given, there are variants of the model that can be used by a strict mother as well. There are many mothers, especially tough single mothers, who function as strict fathers. But the model is an idealization, and is intended here only as that. I believe it is a cognitively real idealized model, that is, a model that Americans grow up knowing implicitly. There are variations on it and I will discuss some of them below.

The Strict Father model presupposes a folk theory of human nature that I will call "folk behaviorism":

People, left to their own devices, tend simply to satisfy their desires. But, people will make themselves do things they don't want to do in order to get rewards; they will refrain from doing things they do want to do in order to avoid punishment.

This is used in the Strict Father model on the assumption that punishment for violating strict moral rules and praise for following them will result in the child's learning to obey those rules. The entire Strict Father model is based on the further assumption that the exercise of authority is itself moral; that is, it is moral to reward obedience and punish disobedience. I will refer to this most basic assumption as the Morality of Reward and Punishment.

Reward and punishment are moral not just for their own sake. They have a further purpose. The model assumes that life is struggle for survival. Survival in the world is a matter of competing successfully. To do so, children must learn discipline and build character. People are disciplined (punished) in order to become self-disciplined. The way self-discipline is learned and character is built is through obedience. Being an adult means that you have become sufficiently self-disciplined so that you can be your own authority. Obedience to authority thus does not disappear. Being self-disciplined is being obedient to your own authority, that is, being able to carry out the plans you make and the commitments you undertake. That is the kind of person you are supposed to be, and the Strict Father model of the family exists to ensure that a child becomes such a person.

There is also a pragmatic rationale for creating such people. It is that the world is difficult and people have to be self-disciplined to be able to survive in a difficult world. Rewards and punishments by the parent are thus moral because they help to ensure that the child will be able to survive on its own. Rewards and punishments thus benefit the child, which is why punishment for disobedience is understood as a form of love.

According to this model, if you are obedient, you will become self-disciplined, and only if you are self-disciplined can you succeed. Success is therefore a sign of having been obedient and having become self-disciplined. Success is a just reward for acting within this moral system. This makes success moral.

Competition is a crucial ingredient in such a moral system. It is through competition that we discover who is moral, that is, who has been properly self-disciplined and therefore deserves success, and who is fit enough to survive and even thrive in a difficult world.

Rewards given to those who have not earned them through competition are thus immoral. They violate the entire system. They remove the incentive to become self-disciplined and they remove the need for obedience to authority.

But this model, as we observed above, is only partly a prescription for enabling children to survive and thrive in a difficult world. It is a model about what a person should be—self-disciplined enough to make his own plans, undertake his own commitments, and carry them out.

But if a person is to be this way, the world must be a certain way too. The world must be and must remain a competitive place. Without competition, there is no source of reward for self-discipline, no motivation to become the right kind of person. If competition were removed, self-discipline would cease and people would cease to develop and use their talents. The individual's authority over himself would decay. People would no longer be able to make plans, undertake commitments, and carry them out.

Competition therefore is moral; it is a condition for the development and sustenance of the right kind of person. Correspondingly, constraints on competition are immoral; they inhibit the development and sustenance of the right kind of person.

Even if survival were not an issue, even if the world could be made easier, even if there were a world of plenty with more than enough for everybody, it would still not be true that parceling out a comfortable amount for everyone would make the world better and people better. Doing that would remove the incentive to become and remain self-disciplined. Without the incentive of reward and punishment, self-discipline would disappear, and people would no longer be able to make plans, undertake commitments, and carry them out. All social life would come to a grinding halt. To prevent this, competition and authority must be maintained no matter how much material largesse we produce.

If competition is a necessary state in a moral world—necessary for producing the right kind of people—then what kind of a world is a moral world? It is necessarily one in which some people are better off than others, and they deserve to be. It is a meritocracy. It is hierarchical, and the hierarchy is moral. In this hierarchy, some people have authority over others and their authority is legitimate.

Moreover, legitimate authority imposes responsibility.

Just as the strict father has a duty to support and protect his family, so those who have risen to the top have a responsibility to exercise their legitimate authority for the benefit of all under their authority. This means:

1. Maintaining order; that is, sustaining and defending the system of authority itself.
2. Using that authority for the protection of those under one's authority.
3. Working for the benefit of those under one's authority, especially helping them through proper discipline to become the right kind of people.
4. Exercising one's authority to help create more self-disciplined people, that is, the right kind of people, for their own benefit, for the benefit of others, and because it is the right thing to do.

In short, this model of the family comes with an idea of what the right kind of person is and what kind of world will produce and sustain such people.

This model of the family does not occur alone and isolated in one's conceptual system. To accept this model of the family is also to accept implicitly certain moral priorities that naturally go with it, many of which are metaphorical in nature. These moral priorities are directly expressed in priorities given to certain metaphors we all have in our conceptual systems. Such a set of moral priorities, together with the above vision of what a person should be and what the world should be like, is what I will call Strict Father morality.

The metaphor analysis that I am about to give is based on contemporary metaphor theory within cognitive linguistics and more broadly within cognitive science. It is worth repeating, before I begin, that the analysis of a concept as metaphorical does not in itself either impugn or confirm its validity. It is simply a technical recognition of the nature of the concept and the way that it functions in our conceptual systems. Here are the metaphors that have highest priority in Strict Father morality:

MORAL STRENGTH

The metaphor that is central to Strict Father morality is the metaphor of Moral Strength. This is a complex metaphor with a number of parts, beginning with:

- Being Good Is Being Upright.
- Being Bad Is Being Low.

Examples include sentences like:

> He's an *upstanding* citizen. He's on the *up and up*. That was a *low* thing to do. He's *underhanded*. He's a *snake* in the grass.

Doing evil is therefore moving from a position of morality (uprightness) to a position of immorality (being low). Hence,

- Doing Evil Is Falling.

The most famous example, of course, is the *fall* from grace.

A major part of the Moral Strength metaphor has to do with the conception of immorality, or evil. Evil is reified as a force, either internal or external, that can make you fall, that is, commit immoral acts.

- Evil Is a Force (either internal or external).

Thus, to remain upright, one must be strong enough to "stand up to evil." Hence, morality is conceptualized as strength, as having the *moral fiber* or *backbone* to resist evil.

- Morality Is Strength.

But people are not simply born strong. Moral strength must be built. Just as in building physical strength, where self-discipline and self-denial ("no pain, no gain") are crucial, so moral strength is also built through self-discipline and self-denial, in two ways:

1. Through sufficient self-discipline to meet one's responsibilities and face existing hardships;
2. Actively through self-denial and further self-discipline.

To summarize, the metaphor of Moral Strength is a set of correspondences between the moral and physical domains:

- Being Good Is Being Upright.
- Being Bad Is Being Low.
- Doing Evil Is Falling.
- Evil Is a Force (either internal or external).
- Morality Is Strength.

One consequence of this metaphor is that punishment can be good for you, since going through hardships builds moral strength. Hence, the homily "Spare the rod and spoil the child." By the logic of this metaphor, moral weakness is in itself a form of immorality. The reasoning goes like this: A morally weak person is likely to fall, to give in to evil, to perform immoral acts, and thus to become part of the forces of evil. Moral weakness is thus nascent immorality, immorality waiting to happen.

There are two forms of moral strength, depending on whether the evil to be faced is external or internal. *Courage* is the strength to stand up to *external* evils and to overcome fear and hardship.

Much of the metaphor of Moral Strength is concerned with *internal* evils, cases where the issue of self-control arises. What has to be strengthened is one's will. One must develop will power in order to exercise control over the body, which is seen as the seat of passion and desire. Desires—typically for money, sex, food, comfort, glory, and things other people have—are seen in this metaphor as "temptations," evils that threaten to overcome one's self-control. Anger is seen as another internal evil to be overcome, since it too is a threat to self-control. The opposite of self-control is "self-indulgence," a concept that makes sense only if one accepts the metaphor of moral strength. Self-indulgence is seen in this metaphor as a vice, while frugality and self-denial are virtues. The list of the seven deadly sins is a catalogue of internal evils to be overcome: greed, lust, gluttony, sloth, pride, envy, and anger. It is the metaphor of Moral Strength that makes them sinful. If we had no metaphor of Morality As Strength, there would be no sinfulness in any of these. The corresponding virtues are charity, chastity, temperance, industry, modesty, satisfaction with one's lot, and calmness. It is the metaphor of Moral Strength that makes these "virtues."

This metaphor has an important set of entailments:

- The world is divided into good and evil.
- To remain good in the face of evil (to "stand up to" evil), one must be morally strong.
- One becomes morally strong through self-discipline and self-denial.
- Someone who is morally weak cannot stand up to evil and so will eventually commit evil.
- Therefore, moral weakness is a form of immorality.
- Self-indulgence (the refusal to engage in self-denial) and lack of self-control (the lack of self-discipline) are therefore forms of immorality.

Moral Strength thus has two very different aspects. First, it is required if one is to stand up to some externally defined evil. Second, it itself defines a form of evil, namely, the lack of self-discipline and the refusal to engage in self-denial. That is, the metaphor of Moral Strength defines forms of internal evil.

Those who give a very high priority to Moral Strength see

it, of course, as a form of idealism. The metaphor of Moral Strength sees the world in terms of a war of good against the forces of evil, which must be fought ruthlessly. Ruthless behavior in the name of the good fight is thus seen as justified. Moreover, the metaphor entails that one cannot respect the views of one's adversary: evil does not deserve respect, it deserves to be attacked!

The metaphor of Moral Strength thus imposes a strict us-them moral dichotomy. It reifies evil as the force that moral strength is needed to counter. Evil must be fought. You do not empathize with evil, nor do you accord evil some truth of its own. You just fight it.

Moral strength, importantly, imposes a form of asceticism. To be morally strong you must be self-disciplined and self-denying. Otherwise you are self-indulgent, and such moral flabbiness ultimately helps the forces of evil.

In Strict Father morality, the metaphor of Moral Strength has the highest priority. Moral Strength is what the strict father must have if he is to support, protect, and guide his family. And it is a virtue that he must impart to his children if they are to become self-disciplined and self-reliant.

The metaphor of Moral Strength provides a mode of reasoning. Anything that promotes moral weakness is immoral. If welfare is seen as taking away the incentive to work and thus promoting sloth, then according to the metaphor of Moral Strength, welfare is immoral. What about providing condoms to high school students and clean needles to intravenous drug users to lower teenage pregnancy and stop the spread of AIDS? The metaphor of Moral Strength tells us that teenage sex and illegal drug use result from moral weakness—a lack of self-control—and therefore they are immoral. Providing condoms and clean needles accepts that immorality, and that, according to Moral Strength, is also a form of evil. A morally strong person should be able to "Just say no" to sex and drugs. Anyone who can't is morally weak, which is a form of immorality, and immoral people deserve punishment. If you unconsciously reason according to the metaphor of Moral Strength, then all this is just common sense.

An important consequence of giving highest priority to the metaphor of Moral Strength is that it rules out any explanations in terms of social forces or social class. If moral people always have the discipline to just say no to drugs or sex and to support themselves in this land of opportunity, then failure to do so is moral weakness, and hence immorality. If the metaphor of Moral Strength has priority over other forms of explanation, then your poverty or your drug habit or your illegitimate children can be explained only as moral weakness, and any discussion of social causes cannot be relevant.

It should be clear from this discussion why Moral Strength is an instance of metaphorical thought. Good people are not literally upright. Becoming immoral is not literally falling. Evil is not literally a force that can make an upright person fall. Morality is not literally the physical strength to stand up to a force. Words like *upright, fall, backbone, stand up to,* and so on are taken from the physical domain and applied to morality by this metaphor.

The metaphorical view of morality as strength is a product of the human mind. But it is not an arbitrary product. It is grounded in a fact about experiential well-being, that it is better to be strong than to be weak. This makes strength a natural metaphor for morality, but the fact that Moral Strength is a natural metaphor does not mean that it is literally true.

Of course, the metaphorical nature of Moral Strength does not invalidate the metaphor. But the fact that it is a product of the human mind should make us look long and hard at it, just as one should look long and hard at any common metaphor for something as important as morality.

One of the most striking entailments of the Moral Strength

metaphor is the following: To build moral strength you have to work at it actively through self-discipline and self-denial. You don't get to be morally strong by just sitting around doing nothing. Since you have to *build* moral strength, that means you don't have it to start with. Therefore, we all start out morally weak, that is, with an overwhelming tendency to do immoral things. Unless our parents intervene to discipline us, we will naturally become immoral.

This is almost, but not quite, an instance of the doctrine of original sin. It does, however, entail a view of children not as naturally good but as naturally tending toward evil unless some strong corrective action is taken. This view of children fits naturally with another important metaphor for morality—Moral Authority—which is also given high priority by the Strict Father model.

MORAL AUTHORITY

Moral authority is patterned metaphorically on parental authority, and so let us begin with the family. The legitimacy of parental authority comes from (1) the inability of the child to know what is in the best interests of himself and the family and to act in those best interests, (2) the parent's having the best interests of the child and the family at heart and his acting on those best interests, (3) the ability of the parent to know what is best for the child, and (4) the social recognition that the parent has responsibility for the well-being of the child and the family.

Within the Strict Father model, the parent (typically the father) sets standards of behavior and punishes the child if the standards are not met. Moral behavior by the child is obedience to the parent's authority. But just as importantly, the exertion of authority is moral behavior on the part of the parent, and it is immoral for the parent to fail to exert authority, that is, to fail to set standards of behavior and to enforce them through punishment. The reason for this is the belief that punishing disobedient children will deter disobedience; that is, it will make children behave morally.

In short, good parents set standards, good children obey their parents, disobedient children are bad children, good parents punish disobedient children, punishment makes disobedient (bad) children into obedient (good) children, and parents who don't punish are bad parents because they produce bad children by not punishing them when they disobey.

In general, the concept of moral authority within communities is patterned on parental authority within families. The general metaphor looks like the following:

- A Community Is a Family.
- Moral Authority Is Parental Authority.
- An Authority Figure Is a Parent.
- A Person Subject to Moral Authority Is a Child.
- Moral Behavior by Someone Subject to Authority Is Obedience.
- Moral Behavior by Someone in Authority Is Setting Standards and Enforcing Them.

This metaphor takes the special case of parental authority and generalizes it to all moral authority. Metaphors like this that characterize a general case in terms of a special case are called "Generic-Is-Specific" metaphors. (See References, A1: Lakoff and Turner 1989.)

The Strict Father model of the family comes with a model of parental authority, the one given above. The metaphor of Moral Authority generalizes that model to all forms of moral authority. Applying this metaphor to legitimacy conditions for parental authority, we arrive at legitimacy conditions for all forms of moral authority:

The legitimacy of moral authority comes from (1) the inability of the person subject to moral authority to

know what is in the best interests of himself and the
community and to act in those best interests; (2) the
authority figure's having the best interests of the community and the person subject to authority at heart
and acting on those best interests; (3) the ability of
the authority figure to know what is best for the community and the person subject to authority; and
(4) the social recognition that the authority figure has
responsibility for the well-being of the community
and the person subject to authority.

Since the Strict Father model puts forth a particular model
of parental authority, it implies a corresponding model of
moral authority in general via this metaphor.

The authority figure sets standards of behavior and
punishes those subject to authority if the standards are
not met. Moral behavior by someone subject to authority is obedience to the authority figure. But just as importantly, the exertion of authority is moral behavior
on the part of the authority figure, and it is immoral
for the authority figure to fail to exert authority, that
is, to fail to set standards of behavior and to enforce
them through punishment.

This is the Strict Father version of moral authority, and one
can see it applied in many arenas of life where moral authority is an issue and institutions are patterned on the idea of
moral authority: athletic teams, the military, law enforcement, business, religion, and so on. As we shall see, there
is also a Nurturant Parent version of moral authority which
is very different.

The Resentment Toward "Illegitimate" Moral Authority

The conditions for the legitimacy of parental authority play
an important role in Strict Father morality, since the Moral

Authority metaphor turns them into conditions on legitimate moral authority in general. The crucial conditions are these: (1) A parent must know better than the child what the child's and the family's best interests are. (2) The parent must be acting in those best interests. These conditions cease to hold as the child becomes mature. At maturity, a child is assumed to be able to determine and act on his best interests for himself. A "meddling" parent is one who asserts his authority in the child's life when he has no business doing so, when his child is mature enough to have authority over his own life. In the Strict Father model, the father must know when his authority ends, after which any illegitimate intrusion by him is resented mightily.

When the Moral Authority metaphor transfers these conditions from parents to general authority figures, it also creates conditions for illegitimate moral authority and for resentment against it: This happens (1) when the person subject to authority knows better than the authority figure what his and the community's best interests are and is capable of acting in those interests; (2) when the authority figure is not acting in the best interests of the person subject to authority and of the community.

Advocates of Strict Father morality show such a resentment of illegitimate authority, not just toward meddling parents but toward any moral authority seen to be illegitimately meddling in their lives. The federal government is a common target. We regularly hear arguments that the federal government doesn't know what's best for people, that people know what's best for themselves, and that the government is not acting in the interests of ordinary people. Therefore, federal authority should be shifted to local governments or eliminated altogether.

It is important to understand that the resentment toward authority that is perceived to be illegitimate does not in any way contradict the central role of legitimate moral authority in Strict Father morality. Rather, it is a consequence of the conditions on the legitimacy of parental and, therefore, general moral authority.

These conditions on the legitimacy of moral authority come, in part, out of certain peculiarities of the American Strict Father model of the family. In other cultures, which have their own versions of Strict Father families, it is not always the case that the father's legitimate authority ends when the child reaches maturity. China is a case in point. In many cultures, it is not the case that children are expected to become fully self-reliant and go off on their own at maturity. Examples include Italy, France, Spain, and Israel, as well as China. Correspondingly, such cultures do not show the same deep resentment toward meddling parents in their versions of Strict Father families. And as we shall see below, they do not show the same resentment toward governmental authority.

Strict Father morality has a form of resentment toward the meddling and intrusion of "illegitimate" authority figures that appears not to be traditional in Western culture, but rather seems to be an American innovation—a consequence of the peculiarly American version of the Strict Father family. Strict Father morality is sometimes mistakenly called "traditional morality," and it is important to understand that aspects of it are not traditional at all but recent innovations, especially the idea that mature children are on their own and parents are not to meddle.

Retribution

Strict Father morality makes a choice among the schemas characterized by the Moral Accounting metaphor. Strict Father morality requires retribution rather than restitution for harming someone or for violations of moral authority. One would expect those who have Strict Father morality to favor the death penalty. They choose balancing the moral books (a death for a death) over preserving life for its own sake. One would expect advocates of Strict Father morality to want prison sentences to be harsher and prison life meaner. One would also expect them to believe, in accord with Moral Authority, that strict punishment of criminal offenders will deter crime.

MORAL ORDER

The metaphor of Moral Order fits naturally with the metaphor of Moral Authority, as well as with the literal parental authority central to the Strict Father family. This metaphor is based on a folk theory of the natural order: The natural order is the order of dominance that occurs in the world. Examples of the natural order are as follows:

God is naturally more powerful than people.
People are naturally more powerful than animals and plants and natural objects.
Adults are naturally more powerful than children.
Men are naturally more powerful than women.

The metaphor of Moral Order sees this natural hierarchy of power as moral. The metaphor can be stated simply as:

• The Moral Order Is the Natural Order.

This metaphor transforms the folk hierarchy of "natural" power relations into a hierarchy of moral authority:

God has moral authority over people.
People have moral authority over nature (animals, plants, and natural objects).
Adults have moral authority over children.
Men have moral authority over women.

But this does not merely legitimize power relations, since those in a position of moral authority also have a moral responsibility for the well-being of those they have authority over. Thus, we have as a consequence:

God has a moral responsibility for the well-being of human beings.

Human beings have a responsibility for the well-being of animals, plants, and the rest of nature.

Adults have a responsibility for the well-being of children.

Men have a responsibility for the well-being of women.

The Strict Father model of the family is, in part, a reflection of the moral order, as defined by this version of the metaphor. The father has a moral responsibility to support his wife and children and to regulate their behavior.

The Moral Order metaphor plays a crucial role in an important interpretation of the Judeo-Christian religious tradition. It is an entailment of this metaphor that God cares about human beings in the same way as parents care about their children or shepherds care about their flocks or farmers care about their crops. Logically, after all, there is no reason that a supreme being should care about lesser beings. But if the order of dominance is a moral order, then God does care about mere mortals; setting the rules and enforcing them is how he shows he cares, and in return for his care, we owe him obedience.

The consequences of the metaphor of Moral Order are enormous, even outside religion. It legitimates a certain class of existing power relations as being natural and therefore moral, and thus makes social movements like feminism appear unnatural and therefore counter to the moral order. It legitimates certain views of nature, e.g., nature as a resource for human use and, correspondingly, man as steward over nature. Accordingly, it delegitimizes other views of nature, e.g., those in which nature has inherent value. In addition, it focuses attention on questions of natural superiority, and so stimulates interest in books like *The Bell Curve*. The issue raised by *The Bell Curve* is not just whether it is a practical waste of time and money to try to educate nonwhites. The real issue is virtually unmentionable: whether whites are naturally superior to nonwhites and hence, according to this metaphor, morally superior to nonwhites.

The metaphor of the Moral Order has a long history in Western culture—a history which is, from the perspective of contemporary American liberal values, not very pretty. It is referred to in a more elaborate version as The Great Chain of Being (see References, E, Lovejoy 1936; A1, Lakoff and Turner 1989, chap. 4). In earlier versions, the moral order included the nobility having moral authority over commoners. Nietzsche's moral theory rested on the Moral Order metaphor, especially on the version in which nobility confers moral authority. In Nazi morality, Aryans ranked higher in the moral order than Jews and Gypsies. For white supremacists, whites rank higher in the moral order than nonwhites. For superpatriots, the U.S. ranks higher in the moral order than any other nation in history. And there are people (typically, wealthy people) who believe that the rich are morally superior to the poor. Indeed, that belief is explicit in forms of Calvinism, where worldly goods are a reflection of righteousness.

The idea that the rich have moral authority over the poor fits American Strict Father morality very well. Start with the American Dream, the stereotypic assumption that America is truly a land of opportunity where anyone with self-discipline and talent can, through hard work, climb the ladder of success. It follows that anyone who has been in the country long enough and is not successful has either not worked hard enough or is not talented enough. If he has not worked hard enough, he is slothful and hence morally weak. If he is not talented enough, then he ranks lower than others in the natural order and hence lower in the moral order. The rich (who are disciplined and talented enough and who have worked hard enough to become rich) deserve their wealth and the poor (either through lack of industry or talent) deserve their poverty. The rich are thus not just more powerful than the poor, they also have moral authority over the poor and with it the moral responsibility to tell the poor how to live: build self-discipline, work hard, climb the economic ladder, and so become self-reliant.

MORAL BOUNDARIES

Strict Father morality, with its sharp division between good and evil and its need for the setting of strict standards of behavior, naturally gives priority to the metaphor of Moral Boundaries.

It is common to conceptualize action as a form of self-propelled motion and purposes as destinations that we are trying to reach. Moral action is seen as bounded movement, movement in permissible areas and along permissible paths. Given this, immoral action is seen as motion outside of the permissible range, as straying from a prescribed path or transgressing prescribed boundaries. To characterize morally permissible actions is to lay out paths and areas where one can move freely. To characterize immoral action is to limit one's range of movement. In this metaphor, immoral behavior is "deviant" behavior, a form of metaphorical motion into unsanctioned areas, along unsanctioned paths, and toward unsanctioned destinations.

Because human purposes are conceptualized in terms of destinations, this metaphor has considerable consequences. Since action is self-propelled motion in this metaphor, and such motion is always under the control of whoever is moving, it follows that any destination is a freely chosen destination and that the destinations chosen by others have been rejected. Someone who moves off of sanctioned paths or out of sanctioned territory is doing more than merely acting immorally. He is rejecting the purposes, the goals, the very mode of life of the society he is in. In doing so, he is calling into question the purposes that govern most people's everyday lives. Such "deviation" from social norms goes beyond mere immorality. Actions characterized metaphorically as "deviant" threaten the very identity of normal people, calling their most common and therefore most sacred values into question.

But "deviant" actions are even more threatening than that. Part of the logic of this metaphor has to do with the effect of deviant behavior on other people. Metaphorically, someone who deviates from a tried and true path is creating a new path that others will feel safe to travel on. Hence, those who transgress boundaries or deviate from a prescribed path may "lead others astray" by going off in a new direction and creating a new path.

The Moral Boundaries metaphor thus interacts powerfully with one of the most important metaphors in our conceptual system: Life Is a Journey. Choosing a particular path, a "direction" in your life, can affect the whole rest of your life. Imagine a parent who says, "Our son left the church; I can't understand why he turned his back on our way of life like that." The paths you choose can be life paths, and if morality is seen as going along a particular path, then deviating from that path can be seen as entering an immoral way of life. It is for this reason that the very idea of "deviance" is so powerful. In creating new paths, the "deviant" can make those paths appear safe to others and thus lead them to change their lives.

Thus, the actions of people who are "deviant" have effects far beyond themselves. Their acts call into question traditional moral values and traditional ways of leading a moral life, and they may make the "deviant" way seem safe, normal, and attractive. If someone smokes marijuana, has no ill effects, and leads a happier, less stressed life, then he has forged a path that others who know him will feel safe going on. If a young woman has sex out of wedlock, has no ill effects, and goes on to have a happy life, then those who know her may feel safe taking such a path.

People who "deviate" from the tried and true path arouse enormous anger because they threaten the identities of those who follow traditional "straight and narrow" moral paths, but also because they are seen as threats to the community. For the protection of the community, they need to be isolated and made outcasts.

CONSTRAINTS ON FREEDOM

Since freedom of action is understood metaphorically as freedom of motion, moral boundaries can be, and often are, seen as constraints on freedom. For this reason, people who want to impose their moral views on others are seen as restricting the freedom of others.

RIGHTS AS PATHS

The Moral Boundaries metaphor is also central in the definition of what we mean by rights. A "right" is not only a form of metaphorical credit, as discussed above; it is also metaphorically a clear path along which one can move freely without being impeded. Hence, via the metaphor that action is motion, a right is a right-of-way, a region in which one can act freely without restraint. Since moral bounds leave open some and close off other regions of free movement, they define rights to free action without interference.

Those rights impose a corresponding duty not to limit that freedom of action, and governmental action may be required if that right is to be respected. For example, proponents of unlimited property rights, such as real estate developers, see environmental regulations as restrictions on the free disposition of their property and therefore want to eliminate governmental regulations as a restriction on their rights. On the other hand, people who see human beings as having a right to a clean, healthy, and biologically diverse environment see unregulated development as "encroaching" on *their* rights. Moral and legal boundaries can thus be seen from two perspectives: what is one man's constraint on free movement is another man's protection against encroachment. This is the logic by which moral and legal bounds create conflicts of rights.

MORAL ESSENCE

A central notion in Strict Father morality is "character," which is taken to be a kind of essence that is developed in childhood and then lasts a lifetime. The centrality of character in Strict Father morality gives priority to the general metaphor of Moral Essence, in terms of which the concept of character is defined.

Physical objects are made of substances, and how they behave depends on what they are made of. Wood burns and stone doesn't. Hence, objects made of wood will burn and objects made of stone will not.

We commonly understand people metaphorically as if they were objects made of substances that determine how they will behave. It is thus common to conceive of a person as if he had an essence or a collection of essences that determined his behavior. This is called the Metaphor of Essence:

- A Person Is an Object.
- His Essence Is the Substance the Object
   Is Made Of.

Imagine judging someone to be inherently stubborn or reliable. To do so is to assign that person an inherent trait, an essential property that determines how he will act in certain situations. If the trait is a moral trait, then we have a special case of the

Metaphor of Essence—the metaphor of Moral Essence. In the field of social psychology, there is an expert version of this metaphor called the "trait theory of personality." We are discussing the folk version of that expert theory here.

According to the metaphor of Moral Essence, people are born with, or develop in early life, essential moral properties that stay with them for life. Such properties are called "virtues" if they are moral properties, and "vices" if they are immoral properties. The collection of virtues and vices attributed to a person is called that person's "character." When people say "She has a *heart of gold*" or "He doesn't have *a mean bone in his body*" or "He's *rotten to the core*," they are making use of the metaphor of Moral Essence. That is, they are saying that the person in question has certain essential moral qualities that determine certain kinds of moral or immoral behavior.

To attribute a moral essence to someone is to make a moral judgment about that person in general, not just a judgment about some single act. Sometimes those judgments are absolute, as when we consider someone as inherently good or evil. But such cases are rare. It is much more normal to attribute particular virtues to people, often a complex combination of virtues that define that person's character.

Those moral virtues are themselves defined relative to particular moral schemes, like those we have been discussing. The metaphor of Moral Strength defines virtues like self-discipline, courage, temperance, sobriety, chastity, industry, and perseverance; and vices like self-indulgence, cowardliness, lust, drunkenness, sloth, and faintheartedness. Virtues and vices don't simply exist objectively. What counts as a virtue or a vice depends upon the moral schemes that one gives priority to. As we shall see below, when we discuss Nurturant Parent morality, that moral system gives priority to different virtues such as care, compassion, kindness, social responsibility, tact, open-mindedness, inquisitiveness, and flexibility, as well as different vices, such as selfishness, insensitivity, meanness, social irresponsibility, tactlessness, closed-mindedness, and inflexibility.

The metaphor of Moral Essence has three important entailments:

- If you know how a person has acted, you know what his character is.
- If you know what a person's character is, you know how he will act.
- A person's basic character is formed by adulthood (or perhaps somewhat earlier).

These entailments form the basis for certain currently debated matters of social policy.

Take, for example, the "Three strikes and you're out" law now gaining popularity in the United States. The premise is that repeated past violations of the law indicate a character defect, an inherent propensity to illegal behavior that will lead to future crimes. Since the felon's basic character is formed by adulthood, he is "rotten to the core" and cannot change or be rehabilitated. He therefore will keep performing crimes of the same kind if he is allowed to go free. To protect the public from his future crimes, he must be locked up for life—or at least a very long time.

Or take the proposal to take illegitimate children away from impoverished teenage mothers and put them in orphanages or foster homes. The assumption is that the mother is immoral, that it is too late to change her since her character is already formed. If the child stays with the mother, he will also develop an immoral character. But if the child is removed from the mother before his character is formed, the child's character can be shaped in a better way.

The metaphor of Moral Essence is a significant part of our moral repertoire. It resides deep in our conceptual systems. It is used to define virtues and vices of all sorts. It plays a role in our political life, and it is used by liberals and conservatives alike. But it is given a high priority in Strict Father morality because of the importance of discipline to character development in the Strict Father model of the family.

MORAL WHOLENESS

In the Strict Father model of the family, the father is the parental authority who sets strict rules for what counts as right and wrong. Correspondingly, the metaphor of Moral Strength sees evil as a force in the world and therefore sees a strict demarcation between good and evil. The metaphor of Moral Boundaries conceptualizes moral and immoral action spatially by strict boundaries and clearly delineates paths of behavior. Those who engage in deviant behavior, who deviate from those paths and transgress those boundaries, are thus threats to society since they blur the established boundaries between morality and immorality. Strict Father morality requires that there are natural, strict, uniform, unchanging standards of behavior that must be followed if society is to function.

Another way to conceptualize uniform standards of behavior is through the metaphor of Moral Wholeness. Wholeness entails a homogeneity—things made of radically different substances may not hold together. Wholeness also entails an overall unity of form that makes an entity strong and resistant to pressures. Homogeneity and unity of form also make an entity stable and predictable in the way it functions. An object with physical integrity can be trusted to function the way it is supposed to function. Wholeness also entails naturalness—something that has the form that it is supposed to have. When an object that is whole starts to crumble, tear, or rot, it is in danger of not holding together and therefore not being able to function.

Advocates of Strict Father morality speak of "degenerate" people, moral "decay," the "erosion" of moral standards, the "rupture" or "tearing" of our moral fabric, the "chipping away" at, and "crumbling" of, moral foundations. All of these are cases where morality is seen as wholeness and immorality as a departure from that state. Wholeness here is abstract and may apply to any kind of entity: a building can crumble, a hillside erode, an organism decay, a fabric tear, a stone can be chipped away at, and so on. It is the wholeness that is at issue in this metaphor, not whether the entity is a building or a hillside or an organism. The kind of entity that is whole or not doesn't matter. Buildings and hillsides are mere special cases of entities that can be whole or not.

Integrity

Moral Wholeness combines with Moral Essence to yield the virtue of integrity—the virtue of being morally whole. Someone who has integrity has moral wholeness, the moral equivalent of physical wholeness. A person with integrity has consistent moral principles, the moral equivalent of physical homogeneity and parts that form a unified whole. The overall unity of moral principle makes someone with integrity strong—not able to be easily swayed by social or political pressures or fashions. A person with integrity acts predictably, in a way consistent with his moral principles, and can be trusted to act in the way he is morally supposed to act. A person with integrity also acts according to his nature; there is nothing artificial or contrived about him.

Consequences

The metaphor of Moral Wholeness can be stated simply:

- Morality Is Wholeness.
- Immorality Is Degeneration.

The entailments of this metaphorical mode of thought are quite considerable: Moral standards that change with time, or social situation, or ethnicity are a danger to the functioning of society. There is no such thing as progress in morality; what is and is not moral is fixed for all time, and any change of standards in the name of would-be moral progress is really an evil, a chipping away at our moral foundations, a tearing of our moral fabric, and so on. And, above all, it is important to constantly be on the lookout for signs of moral decay and erosion and to stop them immediately, because once rot sets in or the foundation crumbles, repair may be impossible, immorality will become rampant, and society will be unable to function in its natural moral way. Moral decay is therefore so dangerous that one must be constantly on the lookout for it and it must be stopped as soon as possible or it will go too far and be irreversible.

MORAL PURITY

Integrity and the metaphor of Moral Wholeness go hand-in-hand with the metaphor of Moral Purity. Just as homogeneous moral standards are threatened by any lack of homogeneity, so the purity of moral standards is threatened by any impurity. A rotten apple spoils the barrel.

Morality is therefore conceptualized as purity and immorality as impurity, as something disgusting or dirty. Linguistic examples make this clear: That was a *disgusting* thing to do. He's a *dirty* old man. We've got to protect our children from such *filth*. She's as *pure* as the driven snow. We're going to *clean up* this town.

The metaphor can be stated simply as:

- Morality Is Purity.
- Immorality Is Impurity.

The entailments of this metaphor are powerful: Just as physical impurities can ruin a substance, so moral impurities can ruin a person or a society. Just as substances, to be usable, must be purged of impurities, so societies, to be viable, must be purged of corrupting individuals or practices. Immorality can ruin a society and therefore cannot be tolerated. Moral Purity is often paired with Moral Essence. Something that has been "corrupted" is something that has been made impure and hence unusable—such as corrupted blood samples or corrupted databases. Metaphorically, someone who is "corrupt" has an impure essence, which, by Moral Purity and Moral Essence, makes him inherently immoral. Such people must be isolated and removed from the rest of society so that their corrupting effect can be nullified.

MORAL HEALTH

In this culture, impurities are seen as causes of illness. This link between impurity and health has led to a further metaphor in which morality is conceptualized as health and immorality as disease.

- Morality Is Health.
- Immorality Is Disease.

This leads us to speak of immoral people as "sick" or having "a diseased mind." And it leads one to speak of the spread of immoral behavior as "moral contagion," and of sudden unexpected immoral behavior on a large scale as an "outbreak" of immorality.

The logic of this metaphor is extremely important: Since diseases can spread through contact, it follows from the metaphor that immorality can spread through contact. Hence, immoral people must be kept away from moral people, lest they become immoral too. This is part of the logic behind urban flight, segregated neighborhoods, and strong sentencing guidelines even for nonviolent offenders. The same logic lies behind guilt-by-association arguments: If you are in contact with immoral people, you become immoral.

MORAL SELF-INTEREST

In the Strict Father model of the family, people become self-reliant by using their self-discipline to pursue their self-interest. The pursuit of self-interest is thus moral, providing, of course, that other, "higher" principles like moral authority and moral strength are not violated. Indeed, without the morality of pursuit of self-interest, there would be no moral link between self-discipline and self-reliance.

Moral Self-Interest, as used in the Strict Father model, is a metaphorical version of an economic idea. It is based on a folk version of Adam Smith's economics: If each person seeks to maximize his own wealth, then by an invisible hand, the wealth of all will be maximized. Applying the common metaphor that Well-Being Is Wealth to this folk version of free-market economics, we get: If each person tries to maximize his own well-being (or self-interest), the well-being of all will be maximized. Thus, seeking one's own self-interest is actually a positive, moral act, one that contributes to the well-being of all.

Correspondingly, interfering with the pursuit of self-interest is seen in this metaphor as immoral, since it does not permit the maximization of the well-being of all. In addition, it interferes with the functioning of the Strict Family model, which depends on the assumption that self-discipline will lead to self-reliance. Without this assumption, the discipline imparted by the father to the child will ultimately not help the child to make a living or to satisfy his long-range goals. But if the child is not helped by the discipline imparted by the father, the very legitimacy of the father's authority is called into question. The very legitimacy of the father's authority thus depends on an external condition, the unimpeded path from self-discipline and hard work to self-reliance.

Since the Strict Father model is what holds Strict Father morality together, interference with the pursuit of self-interest threatens the foundations of the whole Strict Father moral framework—from the efficacy of moral strength to the validity of the moral order.

The link between Moral Self-Interest and free-market economics has, of course, not been lost on advocates of Strict Father morality. Controlled-market economies, whether socialist or communist, impede the pursuit of financial self-interest. For this reason, advocates of Strict Father morality have seen socialism and communism as immoral. Not just impractical, but immoral!

Therefore, proposals for the public good that interfere with the pursuit of financial self-interest are commonly seen as immoral by advocates of Strict Father morality. The "do-gooders" are seen as restricting freedom and posing a threat to the moral order. And indeed they are, according to the logic of Strict Father morality.

But Strict Father morality does not make the pursuit of self-interest a good above all other goods. Moral Self-Interest is limited by the rest of the system. For example, it is common for good Strict Fathers to pursue a less lucrative career so that they can spend more time with their families, making sure that their kids grow up properly, that is, self-disciplined, obedient, with good character, following moral precepts, with a proper respect for legitimate authority, and with sufficient nurturance without being spoiled. Moreover, Strict Father morality dictates that many forms of pursuing self-interest are immoral: becoming a drug dealer, luring girls into prostitution, theft, and so on.

Though Strict Father morality in its American form tends to support laissez-faire capitalism, it does have a long history of constraining how capitalism is to function. Business is not to be directly and overtly immoral, to engage in drug-dealing, prostitution, theft, and so on. Business is supposed to show compassion, for example, to be involved in local charities, to help in disaster relief. Business is supposed to promote wholesome community activities, to sponsor Little League teams, bowling leagues, and the like. Business is supposed to be involved in policing itself for the public interest, say, through Better Business Bureaus and professional associations. In short, there is a long history in America of Strict Father morality placing moral constraints on capitalism. There may be a legitimate question of how strong or meaningful these constraints have been, but they are traditional and have been a hallmark of American business for a long time. Because they accord with Strict Father morality, such constraints, which function for the public good, have never been attacked as immoral constraints on free market capitalism.

NURTURANCE IN THE STRICT FATHER SYSTEM

As we shall see below in great detail, there is a conception of moral action as helping helpless people that is conceptualized as akin to the nurturance of young children. In the Strict Father family, children are of course to be nurtured. But nurturance in the Strict Father family takes a somewhat different form than it does outside such a family. Parental authority must be maintained above all, since it is conceived of as the basis of respect for all forms of legitimate authority as well as the basis for learning to exert authority and hence to be self-reliant later in life. Where there appears to be a choice between parental authority and nurturance, parental authority is to be maintained through punishment. But this is not conceptualized as choosing authority over nurturance. Instead, punishment is, in itself, conceptualized as a form of nurturance, because it is seen as teaching self-reliance and respect for legitimate authority. This is "tough love," where punishing children for disobedience shows that you love them.

In a properly functioning Strict Father family, children should learn to abide by parental authority and to become self-disciplined from birth. In such a family, there should be appropriate development of self-discipline and little or no challenging of parental authority. When the Strict Father family is functioning properly, there should be abundant nurturance and little need for punishment.

The metaphor of Morality As Nurturance extends the logic of family-based nurturance to the general domain of help for others in society. Adherents of Strict Father morality are well-known for going to great lengths to help others in their communities who are afflicted by some external disaster: floods, fires, earthquakes, explosions, epidemics, etc. But the same willingness to help does not always extend to those who are seen as irresponsible, or responsible for their own misfortune, or who, if they were sufficiently self-disciplined, should be able to help themselves. In such cases, Strict Father morality may dictate *not* helping for the following reason: *People should accept the consequences of their own irresponsibility or lack of self-discipline, since they will never become responsible and self-disciplined if they don't have to face those consequences.* In such a case, helping would be immoral, since it would encourage moral weakness. An exception would be someone who, through help, will straighten out his life and become sufficiently responsible and self-disciplined. Such a person is worthy of compassion and help.

SELF-DEFENSE

Strict Father morality, as we have seen, comes with strict notions of good and evil, right and wrong. The Strict Father moral system itself is right and good; it could not possibly be wrong and still function as a moral system with a strict right-wrong dichotomy. Opponents of the moral system itself are therefore wrong; and if they try to overthrow the moral system, they will be engaging in an immoral act. The moral system itself must be defended above all.

Let us call this the Principle of Self-Defense: It is the moral duty of all adherents of Strict Father morality to defend Strict Father morality above all else.

In the case of Strict Father morality, there is no shortage of opponents to the system—opponents to absolute criteria for right and wrong, opponents to a hierarchical moral order, opponents to free-market economics, opponents to the priority of moral strength, and so on. Many of these opponents happen to be in the academic world, especially in the humanities, and in the art world. By the Principle of Self-Defense, Strict Father morality categorizes them as immoral, and hostility to the National Endowments for the Arts and Humanities is a natural consequence of this principle.

Another natural consequence of the principle of Self-Defense is the antipathy of Strict Father morality toward homosexuality and feminism. Homosexuality undermines the Strict Father model of the family, which has both a father and a mother, with the father having moral authority over the mother, and with this moral order being legitimated by the metaphor that the Moral Order Is the Natural Order. Homosexuality and feminism, which are both seen as violating the natural order and therefore the moral order, become threats to the moral system itself. Artistic and academic traditions that accept homosexuality as natural and advocate feminism are likewise seen as threats to the moral system, for the same reason.

The Structure of the System

Strict Father morality is organized around the Strict Father model of the family. There is a group of metaphors for morality that fit naturally with that model and are given priority by it. Those metaphors for morality have entailments that go far beyond the Strict Father family model. When taken together, the entailments of those metaphors define a well-organized and far-reaching moral system.

Here is a list of the metaphors discussed and the ways that they fit together in the system. By the "central model," I am, of course, referring to the Strict Father model of the family as given above.

MORAL STRENGTH: This spells out the crucial notion of self-discipline as characterized in the family model and extends it to morality in general.

MORAL AUTHORITY: This builds on parental authority in the central model and extends it to morality generally. In the process, it characterizes a notion of legitimate and illegitimate moral authority.

MORAL ORDER: This legitimizes the Strict Father's authority in the family model, and is important in defining in general what counts as "natural" and hence legitimate authority.

MORAL BOUNDARIES: This allows us to apply spatial reasoning to moral structures.

MORAL ESSENCE: This spells out an important part of what is meant by "character" in the family model.

MORAL WHOLENESS: This provides a way to conceptualize the importance of the unity, stability, and homogeneity of morality as assumed in the central model.

MORAL PURITY: This provides us with a way to conceptualize immorality as portrayed in the family model.

MORAL HEALTH: This allows us to conceptualize the effects of immorality as portrayed in the family model.

MORAL SELF-INTEREST: This provides the crucial link between self-discipline and self-reliance in the family model.

MORALITY AS NURTURANCE: This links nurturance in the family model to helping others in society in general.

Each of these metaphors exists independently of the Strict Father model of the family. Most of them are motivated by experiential morality, as it was described in Chapter 3. Many of them appear in diverse cultures around the world. But in American culture, the peculiarly American version of the Strict Father family model organizes these metaphors in a way that may very well not exist in other cultures.

It is important to see how the logic of Strict Father morality is a consequence partly of the logic of the Strict Father family model, but even more a product of the metaphors listed above that turn that model of the family into a general moral system. Here is a list of the metaphors in the system and what they contribute to Strict Father morality.

MORAL STRENGTH: This contributes a great deal—the strict dichotomy between good and evil, the internal evils, asceticism, and the immorality of moral weakness.

MORAL AUTHORITY: This contributes notions of the legitimacy and illegitimacy of moral authority, and transfers the resentment toward meddling parents into resentment against the meddling of other authority figures.

MORAL ORDER: This legitimizes certain traditional hierarchical power relations and, together with Moral Strength, makes it seem reasonable to think that the rich are either morally or naturally superior to the poor.

MORAL BOUNDARIES: This provides a spatial logic of the danger of deviance.

MORAL ESSENCE: This contributes the idea that there exists an essence called "character," that it can be determined by significant past actions, and that it is a reliable indicator of future actions.

MORAL WHOLENESS: This makes moral unity and uniformity a virtue and suggests the imminent and serious danger of any sign of moral nonunity and nonuniformity.

MORAL PURITY: This associates our visceral reactions of disgust and our logic of the corruption of pure substances with the idea that morality must be unified and uniform. MORAL HEALTH: This adds the logic of disease to the logic of immorality and contributes the idea that contact with immoral people is dangerous because the immorality might spread in a rapid and uncontrollable way like an epidemic. MORAL SELF-INTEREST: This adds the idea that seeking one's self-interest is a moral activity and interfering with the seeking of self-interest is immoral. The application of this metaphor is limited by its role in the system. MORALITY AS NURTURANCE: The role of this metaphor in the system is to specify when helping people is moral. Help is never moral when it interferes with the cultivation of self-discipline and responsibility and therefore leads to moral weakness. Since reward and punishment are assumed to be effective in promoting learning, the giving of nurturance as reward and withholding of nurturance in the name of discipline and punishment can serve the moral purpose of teaching self-discipline and responsibility. Nurturance is not unconditional. It must serve the function of authority, strength, and discipline.

MORAL PRIORITIES

Strict Father morality imposes a hierarchical structure on the metaphors we have just discussed. In this hierarchy of metaphors we can see clearly the moral priorities of Strict Father morality. The metaphors with the highest priority form a group: Moral Strength, Moral Authority, Moral Order, Moral Boundaries, Moral Essence, Moral Wholeness, Moral Purity, and Moral Health. Let us call this the Strength Group.

The Strength Group has the highest priority. Moral Self-

Interest, which functions to link self-discipline to self-reliance, has the next highest priority. Moral Nurturance is last, since it functions in the service of the Strength Group plus Moral Self-Interest. That is, the function of nurturance in this model is to promote strength; providing nurturance is to be a reward for obedience and withholding it, a punishment for disobedience. Nurturance is not the highest end, but a means to that end. That gives it the lowest priority. The priority list is:

1. The Strength Group
2. Moral Self-Interest
3. Moral Nurturance

It is important to bear in mind, as we shall see in the next section, that both Strict Father and Nurturant Parent morality make use of the same metaphors, but the metaphors have opposite priorities in the two systems. As we shall see in the next chapter, there are other metaphors that go along with moral nurturance. These too get the lowest priority in Strict Father morality.

The fact that these are not arbitrary metaphors, but are grounded in everyday well-being and in experiential morality, makes it seem that these metaphorical entailments are just common sense—natural, inevitable, and universal. That is why it is important to separate the metaphors out, to examine them, to understand them thoroughly, and to know what each of them contributes to the overall moral system.

Strict Father Morality is a highly elaborate, unified moral system built around a particular concept of family life and extended to all of morality via metaphors for morality. Those metaphors for morality, for the most part, exist independently of the system, are common in other cultures, and occur in other moral systems. It is the way that they are organized in this system that gives them the overall logical and emotional effect that they have.

Parameters of Variation

The model of family-based morality just given is the central member of a radial category of family models and corresponding moral systems. To date, I have identified four parameters that determine variations on this model:

1. Linear scales
2. The pragmatic-idealistic dimension
3. The presence or absence of particular "clauses" in the Moral Order
4. Moral focus

Let us consider these in turn.

Linear Scales

The violation of a rule may be a matter of degree. Did your teenager come home fifteen minutes late or two hours late? Did your eight-year-old leave a few toys on the floor, or was the room a total mess? Some infringements of a rule are minor, others major, and others in-between. Correspondingly, punishments in general can be relatively harsh or lenient. Do you keep your eight-year-old from watching her favorite TV show, do you send her to bed without dinner, do you take her pants down and whip her with a belt until she offers no more resistance, or do you slap her senseless?

Such linear-scale differences are often not merely quantitative, but qualitative. The difference between a lenient parent, a moderately strict parent, an abusive parent, and a criminal may have to do with the degree to which infractions are taken seriously and with the degree of punishment. Sufficient differences of degree can result in differences of kind.

Idealistic vs. Pragmatic

The model discussed above is an idealistic model. The ideals are to promote self-discipline and self-reliance. The pursuit of self-interest is seen as a means by which a self-disciplined person can achieve the goal of self-reliance. However, the means and the end can be reversed. In a pragmatic variant on the model, the goal is to pursue self-interest and the means for pursuing self-interest are self-discipline and self-reliance.

Thus, a pragmatic strict parent may not care about self-discipline and self-reliance for their own sake, but may want his child to be capable of pursuing her self-interest as competently as possible. He may then feel that self-discipline and self-reliance are the best means to that end. An idealistic strict parent may, on the other hand, see self-discipline and self-reliance as moral ends for his child—the really important things in life. The pursuit of self-interest may just be the best means for a self-disciplined person to achieve self-reliance.

THE MORAL ORDER

The metaphor of the Moral Order links dominance to moral authority. This metaphor has a number of variations, depending on which "clauses" are included. The source domain of the metaphor is the domain of worldly power. In that domain various forms of dominance may occur in a society. Each general instance of dominance is represented by a "clause" of the form "A has dominance over B." The Moral Order metaphor projects a dominance hierarchy onto the moral domain, creating a corresponding hierarchy of legitimate moral authority. A particular version of the metaphor maps a particular set of dominance clauses onto a corresponding set of moral authority clauses of the form "A has moral authority over B."

In the central model given above, a set of dominance clauses, namely,

> God has dominance over human beings.
> Human beings have dominance over nature (animals, plants, and natural objects).
> Adults have dominance over children.
> Men have dominance over women.

are mapped onto a corresponding set of moral authority clauses, in particular,

> God has moral authority over human beings.
> Human beings have moral authority over nature (animals, plants, and natural objects).
> Adults have moral authority over children.
> Men have moral authority over women.

One form of variation on this model is that different dominance clauses from the domain of worldly power are mapped onto the domain of morality.

For example, suppose that the dominance clause "Men have dominance over women" is no longer mapped onto "Men have moral authority over women." Then, what one gets is something like a feminist version of Strict Father morality. In this version of the Strict Father model of the family, the father no longer has authority over the mother in the family and both parents set and enforce the rules equally and make decisions equally.

Take another example. Suppose one took the dominance clause "Whites have dominance over nonwhites" from the domain of worldly power and mapped it onto "Whites have moral authority over nonwhites." This would yield a racist version of Strict Father morality, which might not apply within an all-white family, but would apply as a "moral" principle to society in general.

It is important to note that this parameter of variation is highly constrained. There are not all that many general dominance clauses that are taken to be true in our folk models of the domain of worldly power. Consider a somewhat silly example of what cannot be a dominance clause mapped onto the moral domain. We have no general cultural folk model in which people who dislike cinnamon have worldly dominance over people who like cinnamon. Thus, there cannot be a variant of the Moral Order metaphor that produces "People who dislike cinnamon have moral authority over people who like cinnamon." Cinnamon does not happen to be what cultural dominance is about in our culture, and so it cannot be a distinct determinant of moral authority. Race, sex, and religion are, however, very much involved in cultural dominance and so they enter into possible versions of Moral Order. We will discuss such cases below in Chapter 17.

MORAL FOCUS

A given person may find some aspect of a family model or a family-based moral system to be of overriding importance, and so may give it priority over other aspects of the family model or moral system. The term we use for this is "moral focus."

For example, a strict father may be more concerned with maintaining his authority than with his children really becoming self-disciplined and self-reliant. In this case, we will say that he places his primary moral focus on the maintenance of authority. As a result he may set arbitrary rules that have little or nothing to do with developing self-discipline and self-reliance, but simply are there to show who's boss.

Another case is one where a strict father may give primary moral focus to his own self-reliance and therefore give less moral focus to the protection of his family. Such a parent may be unable to ask for help from friends when his family needs help.

As in the case of the moral order, the possibilities for the use of moral focus are limited to aspects of the model. It is not a variation on the model to have a moral focus on chocolate ice cream, since chocolate ice cream is not in, or implied by, the model. One can only focus on—and give priority to—what is in, or implied by, the model in context.

As we shall see when we get to politics, these four parameters of variation give rise to a considerable number of versions of conservatism—not by some random mechanism, but systematically, since they are defined by the structure of the model itself.

6

Nurturant Parent Morality

Let us now turn to a second moral system built around a model of an ideal family: a Nurturant Parent family. Though this model of the family seems to have begun as a woman's model, it has now become widespread in America among both sexes.

THE NURTURANT PARENT MODEL: A family of preferably two parents, but perhaps only one. If two, the parents share household responsibilities.

The primal experience behind this model is one of being cared for and cared about, having one's desires for loving interactions met, living as happily as possible, and deriving meaning from mutual interaction and care.

Children develop best through their positive relationships to others, through their contribution to their community, and through the ways in which they realize their potential and find joy in life. Children become responsible, self-disciplined, and self-reliant through being cared for and respected, and through caring for others. Support and protection are part of nurturance, and they require strength and courage on the part of parents. The obedience of children comes out of their love and respect for their parents, not out of the fear of punishment.

Open, two-way, mutually respectful communication is crucial. If parents' authority is to be legitimate, they must tell children why their decisions serve the cause of protection and nurturance. The questioning of parents by children is positive, since children need to learn why their parents do what they do, since children often have good ideas that should be taken seriously, and since all family members should participate in important decisions. Responsible parents, of course, have to make the ultimate decisions and that must be clear.

Protection is a form of caring, and protection from external dangers takes up a significant part of the nurturant parent's attention. The world is filled with evils that can harm a child, and it is the nurturant parent's duty to ward them off. Crime and drugs are, of course, significant, but so are less obvious dangers: cigarettes, cars without seat belts, dangerous toys, inflammable clothing, pollution, asbestos, lead paint, pesticides in food, diseases, unscrupulous businessmen, and so on. Protection of innocent and helpless children from such evils is a major part of a nurturant parent's job.

The principal goal of nurturance is for children to be fulfilled and happy in their lives and to become nurturant themselves. A fulfilling life is assumed to be, in significant part, a nurturant life, one committed to family and community responsibility. Self-fulfillment and the nurturance of others are seen as inseparable. What children need to learn most is empathy for others, the capacity for nurturance, cooperation, and the maintenance of social ties, which cannot be done without the strength, respect, self-discipline, and self-reliance that comes through being cared for and caring. Raising a child to be fulfilled also requires helping that child develop his or her potential for achievement and enjoyment. That requires respecting the child's own values and allowing the child to explore the range of ideas and options that the world offers.

When children are respected, nurtured, and communicated with from birth, they gradually enter into a lifetime relationship of mutual respect, communication, and caring with their parents.

Though this model is very different from the Strict Father model, it has one very important thing in common with it. They both assume that the system of childrearing will be reproduced in the child. In the Strict Father model, discipline is incorporated into the child to become, by adulthood, self-discipline and the ability to discipline others. In the Nurturant Parent model, nurturance is incorporated into the child to eventually become self-nurturance (the ability to take care of oneself) and the ability to nurture others.

But the mechanism by which this is accomplished is entirely different in this model, which makes different assumptions about the nature of children in particular and human beings in general. The Nurturant Parent model does not assume that children primarily learn through reward and punishment, nor that adults mostly tailor their actions to rewards and punishments.

Instead, it is assumed that children learn through their attachments to their parents—which are, ideally, secure and loving attachments. They learn to be toward others and toward themselves what their parents are to them, and they learn it in two ways. First, they follow the model of their parents' behavior. Second, through being securely attached to their parents, they become attuned to their parents' expectations and try to meet them. If the parents are careful about making their expectations realistic and appropriately challenging—rather than overdemanding or nondemanding—the children will be able to meet them and develop mastery.

The ideal nurturant parents must be, or become, what they want their children to be: basically happy, empathetic, able to take care of themselves, responsible, creative, communicative, and fair. A securely attached child will be motivated to please his parents and to reproduce the parents' qualities. Parents, by being empathetic and attuned to what a child can do at various stages, gradually encourage the child to do things for himself and for the family at large. Children do this not out of fear of punishment or obedience to authority, but out of a desire to display their mastery, please their parents, and gain respect. Children gradually become self-conscious, that is, conscious of whether their behavior is earning the respect of their parents. Parents must show enthusiasm at their children's display of mastery. Parents become respected because they respect their children. Children come to respect themselves and others through this mechanism.

If children are to become nurturing, they must develop a social conscience. To do so, they must become self-conscious. They need to learn honest questioning and sincere probing, to know what is not so nice about themselves and their parents, both to improve themselves and to have a realistic understanding about who they are and what their parents are like. For this reason, nurturant parents encourage questioning, self-examination, and openness. These are all seen as necessary for the development of a self-conscious and socially conscious person.

In these ways, children become the kind of persons their parents want them to be. They learn to take care of themselves, be responsible, enjoy life, develop their potential, meet the needs and expectations of those they love and respect, and become independent-minded. They also learn to empathize with others, develop social ties, become socially responsible, communicate well, respect others, and act fairly toward them. They become self-nurturant and nurturant to others. In short, they become the right kind of people—what you want people to be if you are going to live in the world with them.

Such a person can function in the world because he can develop his talents, take care of himself, and develop strong bonds of mutual affection with and respect for others. He has an inner strength, having developed it naturally in the course of becoming nurturant of himself and others. And most of all, he understands the nature of interdependence. He understands that bonds of affection and earned mutual respect are stronger than bonds of dominance.

What does the world have to be like if people like this are to develop and thrive? The world must be as nurturant as possible and respond positively to nurturance. It must be a world that encourages people to develop their potential and provides help when necessary. And correspondingly, it must be a place where those who are helped feel a responsibility to help others and carry out that responsibility. It must be a world governed maximally by empathy, where the weak who need help get it from the strong. It must be a world governed as much as possible by bonds of affection, respect, and interdependence. Finally, it must be a world in which the nurturance provided to us by the natural environment is recognized, appreciated, and returned. In short, the natural world must be sustained, and we must do everything we can to sustain it.

The Nurturant Parent model thus defines a moral attitude to the world. It is based on assumptions about human nature, how children learn, and what the right kind of person is. If the world is to be a place that is hospitable to the development of such people, then we have a social responsibility to help make it such a place.

That social responsibility begins with the raising of children. It includes a responsibility to avoid what is harmful. Children should not be brought up by a system of rewards and punishments—and especially not by painful corporal punishment. To a child, corporal punishment is a form of violence, and violence begets violence. If children learn that abuse, punishment, and violence are ways to impose authority and command respect, they will reproduce that behavior and the result will be a violent society. Neglect, the depriving of needed nurturance, has an effect like that of abuse; a child not cared for and respected will not respect and care for others. Cooperation should be stressed rather than competition. Fierce competition brings out aggressive behavior, which will then be duplicated in later life. The nonaggressive side of competition is mastery, which is developed naturally through nurturance and encouragement. Cooperation develops an appreciation for interdependence. An appreciation of pleasure and an aesthetic sense should be cultivated, so that one can develop one's capacity for happiness and one's ability to give to others the gift of one's own happiness. Asceticism should be avoided. Self-denial makes one more likely to deny and disapprove of the happiness of others.

Interdependence is a nonhierarchical relationship. To maximize the benefits of interdependence and cooperation, hierarchical relationships should be minimized. Legitimate authority should be a consequence of the ability to nurture—of wisdom, judgment, empathy, and so on; authority should not come out of dominance. These are exactly the opposite of the childrearing practices of the Strict Father model. The world that the nurturant parent seeks to create has exactly the opposite properties.

The Metaphor System for Nurturant Parent Morality

This view of the family, of childrearing, of what the right kind of person is, and of what the world should be like gives priority to a very different set of metaphors for morality than does the Strict Father model. Where the Strict Father model stressed discipline, authority, order, boundaries, homogeneity, purity, and self-interest, the Nurturant Parent model stresses empathy, nurturance, self-nurturance, social ties, fairness, and happiness. These become the priorities of a rather different moral system, in which morality is conceptualized in terms of what is stressed in this family model, namely, empathy, nurturance, and so on. Thus, the metaphors such as Morality As Empathy and Morality As Nurturance are given primacy in this model. Let us begin with Morality As Empathy.

MORALITY AS EMPATHY

Empathy is understood metaphorically as the capacity to project your consciousness into other people so that you can feel what they feel. We can see this in the language of empathy: *I know what it's like to be in your shoes. I know how you feel. I feel for you.* Now we cannot literally project our consciousness into someone else's mind and body, which is why this notion of empathy is metaphorical. However, it is possible, if we work at it, to imagine being someone else. This is what we have to do, as best we can, if we want to act nurturantly toward someone else. Empathy is the basis of a major conception of morality.

- Morality Is Empathy.

The logic of empathy is this: If you really feel what another person feels, and if you want to feel a sense of well-being, then you will want that person to experience a sense of well-being. Therefore, you will act so as to promote a sense of well-being in that person. To conceptualize moral action as fully empathetic action is more than just abiding by the Golden Rule, to do unto others as you would have them do unto you. The Golden Rule does not take into account that others may have different values than you do. Taking morality as empathy requires basing your actions on their values, not yours. This requires a stronger Golden Rule: Do unto others as *they* would have you do unto them.

The strong Golden Rule is, however, not always applicable. Suppose you are a liberal attempting to empathize with a conservative, whose Strict Father views contradict the very kind of empathy you are trying to use. To adopt his values is to undermine any possible success at implementing your values. When the value system as a whole is at stake, the strong Golden Rule may yield a paradox. To obey it is not to obey it. When discussing values that are less than all-inclusive, the strong Golden Rule is not subject to such a paradox.

The very existence of the traditional weaker Golden Rule suggests that empathy comes in both stronger and weaker forms.

Absolute empathy is simply feeling as someone else feels, with no strings attached. But strings commonly are attached. The reason is that we cannot only project our capacity to feel onto someone else, but we can also project our values onto someone else. Many people can only project their capacity to feel onto someone else if they also project their values onto them. Let us call this *egocentric empathy*. In egocentric empathy, you project your capacity to feel onto another person, keeping *your* values. This yields a weak form of the Golden Rule, what might be called the Brass Rule: Do unto others as you would have them do unto you— but only if they share your values!

It is extremely important to distinguish *egocentric empathy* from *absolute empathy plus moral instruction*. Suppose you have a child and you want to teach that child your moral values. Suppose the child comes to reject some or all of those values, yet you still think it is important to try to teach them to him. Under egocentric empathy, you will not empathize with your child unless he adopts your values. Under absolute empathy with moral instruction, you will empathize with your child despite the difference of values—perhaps doing your best to understand his values—while still trying to get him to adopt your values. Both cases arise regularly in family life throughout the country and the difference is all-important.

Another type of empathy is *affordable empathy*. It is the ability of people who are relatively well-off to empathize with people who are less fortunate than they. The logic of affordable empathy is the Wooden Rule: Do unto others as you would have them do unto you—providing you can afford it easily!

Charity, as it is all too often practiced in this country, combines moral accounting with affordable empathy. It is a way of accruing moral credit by giving something of positive value—typically money—to people who are less well-off than you when you can well afford it. Income tax deductions for charitable contributions are interesting in this light; they permit you to accrue real financial credit rather than mere moral credit.

MORALITY AS NURTURANCE

Nurturance presupposes empathy. A child is helpless, it cannot care for itself. It requires someone to care for it, and to care *for* a child adequately, you have to care *about* a child. You have to project your capacity for feeling onto a child accurately enough to have a sense of what that child needs. This not only requires empathy, it requires constant empathy. It also requires, to a significant extent, putting the child's needs before your own, making sacrifices for your child—though not so much that it prevents one from nurturing adequately.

We have just seen that there are a number of forms of empathy—absolute, egocentric, and affordable. Any particular occurrence of empathy may be a pure form of one of these, but it may also be a mixture, with each form incorporated in various degrees. Empathy is rarely simple or straightforward or pure. Since empathy plays a role in nurturance, each of the many forms of empathy defines a corresponding form of nurturance. Thus, there are complexities of nurturance that mirror the complexities of empathy.

Nurturance also involves rights and duties; it inherently involves morality. A child has a right to nurturance and a parent has a responsibility to provide it. A parent who does not adequately nurture a child is thus metaphorically robbing that child of something it has a right to. For a parent to fail to nurture a child is immoral.

In conceiving of morality as nurturance, this notion of family-based morality is projected onto society in general.

The conception of morality as nurturance can be stated as the following conceptual metaphor:

- The Community Is a Family.
- Moral Agents Are Nurturing Parents.
- People Needing Help Are Children Needing Nurturance.
- Moral Action Is Nurturance.

This metaphor has the following entailments, based on what one knows about being nurturant toward children:

- To nurture children, one must have absolute and regular empathy with them.
- To act morally toward people needing help to survive, one must have absolute and regular empathy with them.
- Nurturance may require making sacrifices to care for children.

- Moral action may require making sacrifices to help
  truly needy people.

If one's community is, further, conceptualized as a family,
a further entailment follows from this metaphor:

- Family members have a responsibility to see that
  children in their family are nurtured.
- Community members have a responsibility to see
  that people needing help in their community are
  helped.

These entailments are widespread among Americans. In
times of disaster, help pours forth for community members
who need help to survive. What limits it is the form of
empathy people have and the issue of who counts as a community member. Those with egocentric empathy will help
only those who share their values. Those who define certain needy people as outside their metaphoric family, that is,
their community, will feel no responsibility for helping
them. Consequently, many Americans see enormous differences between neighbors subject to disasters (who share
their values and are clearly community members) and homeless people (who are not perceived as sharing their values
and who are, for the most part, not seen as community
members).

COMPASSION

Incidentally, the term "compassion" has two intimately related senses, defined relative to moral empathy and moral
nurturance. To "feel compassion" is to experience empathy.
To "show compassion" is to act nurturantly on the basis of
compassionate feelings. There are, of course, limited forms
of compassion that result from the limitations that occur on
empathy and on nurturance, as when you limit compassion

to those who share your values and those whom you perceive as community members.

MORAL SELF-NURTURANCE

You can't care for others adequately if you don't care for yourself. An important part of the morality of nurturance is the requirement of self-nurturance, taking care of one's own basic needs: maintaining one's health, making a living, maintaining interpersonal relationships, and so on. The morality of nurturance and self-nurturance can sometimes be in a precarious balance, when the sacrifices needed to nurture others conflict with taking care of oneself.

It is important to distinguish self-nurturance and self-interest. Self-nurturance is necessary for any adequate moral functioning. Self-interest goes considerably further to the satisfaction of desire, most typically the desire for money and power. These are rather different notions. A *selfish* person is one who puts his self-interest ahead of the needs of those he has a duty to nurture or to share with. But someone who simply attends to his most basic needs, who makes self-nurturance a prerequisite to the nurturance of others, is not selfish. Someone who puts the nurturance of others not only ahead of his self-interest but also ahead of his self-nurturance is *selfless*.

Selflessness is not always what it seems. Though we are taught that selflessness defines saintly behavior, the reality can be quite different. First, selflessness, by moral accounting, imposes moral debts upon the people that the selfless person takes care of. Second, the selfless person, in putting the nurturance of others above self-nurturance, may suffer a decline in health or other capacities and may, because of his selflessness, ultimately have to be taken care of himself. This may impose a considerable burden on others—especially the people he has previously taken care of. Thus, selflessness may impose a considerable cost on those one is selfless towards.

For these reasons, the Morality As Nurturance metaphor implies that self-nurturance is a moral necessity.

Morality As Social Nurturance

There are two varieties of moral nurturance, one about individuals and the other about social relations. When disputes arise or when one person acts unfairly or harms another, social ties can be disrupted or broken. If community members are to empathize with one another and be nurturant toward one another, those social ties must be constantly mended and maintained. The link between nurturance and the maintenance of social ties can be stated as follows:

- Moral Agents Are Nurturing parents.
- Social Ties Are Children Needing Care.
- Moral Action Is the Nurturance of Social Ties.

Much of what we know about the nurturance of children then applies, by this metaphor, to social ties, making our attitudes toward social ties conform to our knowledge about nurturance:

- To act morally, one must attend constantly to social ties.
- One may have to make sacrifices to maintain social ties.
- People who can maintain and mend social ties have a duty to do so.
- It is wrong not to maintain and mend social ties.

The morality of social nurturance is by no means the preserve of women. Anyone who is "diplomatic," who sees the primary moral need as working constantly for compromise and the maintenance of community is living according to this metaphor.

It is important to realize that the Social Nurturance metaphor and the Moral Nurturance metaphor may sometimes contradict each other, even though they form a natural pairing. This occurs when you have to maintain social ties with people in your community who do not believe in or operate by the Moral Nurturance metaphor. Compromising with such people for the sake of maintaining social ties may require compromising on moral nurturance.

MORALITY IS HAPPINESS

Nurturance typically requires sacrifice. And so it might seem strange that a concomitant of moral nurturance would be a view of morality as the cultivation of one's own happiness. Yet such a moral scheme is a consequence of Morality As Nurturance. The reasoning goes like this:

> Unhappy people are less likely to be compassionate (empathetic and nurturant) than happy people, since they are not likely to want others to be happier than they are. Therefore, to promote one's own capacity for compassion (empathy and nurturance), one should make oneself as happy as possible—provided one doesn't hurt anyone in the process.

This view of the morality of happiness is intuitively understood and widespread among a great many whose moral system is Nurturant Parent morality. Incidentally, it is a long-standing part of the Buddhist tradition. There is a reason, after all, why the Buddha is smiling.

Moral Happiness is anything but a form of selfishness or crass self-interest, since the prior commitment to compassion—to help and not hurt others—rules that out. In the context of a commitment to empathy and nurturance, making oneself as happy as possible is anything but mere hedonism, since it promotes empathy and nurturance, which are the most profound forms of moral behavior.

Some Americans have adopted a perverted view of Moral Nurturance, a view in which self-sacrifice is required for the sake of others and someone who pursues his own happiness cannot be seen as being sufficiently nurturant. To those for whom morality mainly means self-sacrifice, the very idea of Moral Happiness is foreign. But, interestingly enough, a great many Americans do have an intuitive sense that, in the context of the priority of empathy and nurturance, the cultivation of one's own happiness serves a moral end. The desire that your children be happy does not contradict your desire that they be empathetic and nurturant and socially responsible—at least not within Nurturant Parent morality.

Such an idea does, however, contradict Moral Strength, where self-denial is seen as serving the moral purpose of building moral strength serving a higher authority. To someone who functions within Strict Father morality, the idea of Moral Happiness will seem like self-indulgence. In the Strict Father moral system, happiness is appropriate as a reward for self-discipline and hard work; in that context, it can serve a moral purpose. But in Nurturant Parent morality, the cultivation of one's own happiness can serve a moral purpose in itself.

MORALITY AS SELF-DEVELOPMENT

Nurturant parents want to see their children develop their abilities—not nonnurturant abilities like the ability to torture people or deceive people or take advantage of them, but abilities that serve nurturance. Thus, Morality As Nurturance entails Morality As Self-Development. What counts as self-development is determined by the rest of the moral system;

it is self-development in the cause of increasing empathy, helping others, nurturing social ties, making people happy, or increasing one's capacity for happiness. Thus appropriate forms of self-development might be education, the development of artistic skills, community service, experience in nature, contact with other cultures, meditation, sensitivity-training, and so on.

Nurturance implies empathy, self-nurturance, the nurturance of social ties, the cultivation of happiness, and self-development. When one understands Morality As Nurturance metaphorically, a host of other metaphors are entailed: Morality As Empathy, Moral Self-Nurturance, Morality As the Nurturance of Social Ties, Morality As Happiness, and Morality As Self-Development. The Nurturant Parent model of the family gives this whole category of moral schemes its highest priority.

MORALITY AS FAIR DISTRIBUTION

The Nurturant Parent model requires that children be nurtured equally and that the responsibilities of parenthood be equally shared between spouses. This gives priority to the metaphor of Morality As Fair Distribution. But that, in itself, does not tell us which of the main models of fair distribution is to be chosen in which circumstance. Recall the models of fair distribution:

- Equality of distribution (one child, one cookie)
- Equality of opportunity (one person, one raffle ticket)
- Procedural distribution (playing by the rules determines what you get)
- Rights-based fairness (you get what you have a right to)

- Need-based fairness (the more you need, the more you have a right to)
- Scalar distribution (the more you work, the more you get)
- Contractual distribution (you get what you agree to)
- Equal distribution of responsibility (we share the burden equally)
- Scalar distribution of responsibility (the greater your abilities, the greater your responsibilities)
- Equal distribution of power (one person, one vote)

The Nurturant Parent model of the family includes some of these. Need-based fairness applies to nurturance for children: younger children may need more attention, teenage children more money. Other situations require fairness to be equality of distribution: one child, one cookie. With a combination of older and younger children, there is a scalar distribution of responsibility; more responsibility is expected of older children. Parents, however, share an equal distribution of responsibility and power. Children's games impose procedural fairness.

In the Nurturant Parent family, the conditions of family life determine the forms of fair distribution. Once the metaphor of Morality As Fair Distribution is extended from the family to life in general, the nature of fairness may become less obvious, or may be determined by other principles (as we shall see below). Nonetheless, Morality As Fair Distribution is a cornerstone of Nurturant Parent morality.

Moral Growth

The nurturance of children is in the service of growth. Children do physically grow. Given that morality is conceptualized in terms of verticality—uprightness, high moral principles, etc.—it is hardly strange that we should have the metaphorical notion of moral growth, in which becoming more moral is seen as "growing." Children grow in response to nurturance and exercise. In the metaphor of Morality As Nurturance, nurturing children corresponds metaphorically to helping someone badly in need of help. A natural extension of that metaphor is the metaphor of Moral Growth, where adults are seen as capable of growing morally either through help (which corresponds to nurturance) or work (the adult correlate of exercise). The metaphor of Moral Growth can be stated as follows:

- The Degree of Morality Is Physical Height.
- Moral Growth Is Physical Growth.
- Moral Norms for People Are Physical Height Norms.

Thus, a "moral midget" is someone of low moral character. We can speak of "moral development," by which we mean the stages of moral sensibility that a child goes through as she or he grows up. We can speak of a person's moral growth as being "stunted," and we know what that means, namely, that she or he has not developed normally and has only reached an early stage of development. Project Head Start has a moral component. The idea is to give a young child a growth spurt in moral development, as well as a "head start" on life's journey.

Moral growth is a central idea in religion and law. The idea of repentance presupposes the possibility for moral growth. In law, "showing remorse" is a demonstration of moral growth and grounds for a reduced prison sentence. The idea of moral growth has long been associated more with liberal than with conservative politics. This comes out clearly in the politics of prisons. The concept of rehabilitation is based on the concept of moral growth. The idea is that if prisoners are treated humanely, taught useful skills, encouraged to get an education, allowed to earn furloughs, and provided with a job upon release, they will have a chance to grow morally and become useful citizens. Not that this is guaranteed, by any means. But if prisoners do grow morally, there is no reason to keep them in prison.

THE MORAL STRENGTH TO NURTURE

The Nurturant Parent has to be strong—strong enough to support and protect a child and strong enough to nurture, which is not an enterprise for weaklings. Nurturing children is exhausting work physically, mentally, and emotionally. For a nurturant parent, strength is in the service of nurturance.

The metaphor of Morality As Nurturance therefore requires an appropriate version of the metaphor of Moral Strength, one in which Moral Strength is in service of Moral Nurturance, that is, where Moral Nurturance has the highest priority and Moral Strength contributes to it. But this means that the metaphor of Moral Strength cannot appear in the system of Nurturant Parent morality in the same form in which it appears in Strict Father morality, where it had the highest priority. In the Nurturant Parent system, the Moral Strength metaphor cannot contradict the metaphors of Morality As Empathy, Nurturance, and Happiness. A version of Moral Strength can appear in the system, but only that part of it that is consistent with Morality As Empathy, Nurturance, and Happiness. Let us consider exactly what that means.

The conceptual metaphor of Moral Strength is stated as in the Strict Father system:

- Being Good Is Being Upright.
- Being Bad Is Being Low.
- Evil Is a Force (external or internal).
- Morality Is Strength.

But when set in the Nurturant Parent system of morality and made to serve and be consistent with the metaphor of Morality As Nurturance and all that goes along with it, the entailments of the Moral Strength metaphor change drastically.

Morality As Empathy and Nurturance requires that one empathize with and be nurturant toward people with different values than one's own, including different moral values. This means that one cannot maintain a strict good-evil dichotomy. To be able to see the world through other people's values and truly empathize with them means that you cannot see all people who have different moral values than yours as enemies to be demonized.

There are external evils, dangers, and hardships and one must be strong to confront them to protect oneself and one's family. That strength comes, not through self-denial and the imposition of discipline for discipline's sake, but rather through the regular exercise of nurturance, which takes strength and hence builds strength.

In addition, the notion of internal evils changes radically. The internal evils that are destructive and that must be confronted are those that interfere with empathy, nurturance, self-nurturance, the maintenance of social ties, the realization of one's potential, and so on. Those internal evils, or moral weaknesses, are lack of social responsibility, selfishness, self-righteousness, narrow-mindedness, inability to experience pleasure, aesthetic insensitivity, lack of curiosity, uncommunicativeness, dishonesty, insensitivity to feelings, inconsiderateness, uncooperativeness, meanness, self-centeredness, and lack of self-respect. In Nurturant Parent morality, the virtues to be taught—the moral strengths—are the opposites of the internal evils: social responsibility, generosity, respect for the values of others, open-mindedness, a capacity for pleasure, aesthetic sensitivity, inquisitiveness, ability to communicate, honesty, sensitivity to feelings, considerateness, cooperativeness, kindness, community-mindedness, and self-respect. A person of good character is a person who has these virtues.

To those raised with Strict Father morality, it may not be obvious why these are, respectively, moral weaknesses and strengths—moral flaws and moral virtues. But from the perspective of nurturance, it is clear. Lack of social responsibility, selfishness, insensitivity to feelings, inconsiderateness, meanness, and dishonesty make it hard to abide by the metaphor of Morality As Nurturance. Incuriosity leads to a lack of knowledge, and since knowledge is needed to be successfully nurturant, the lack of curiosity also limits one's ability for successful nurturance. Self-righteousness and self-centeredness make it difficult to abide by the metaphor of Morality As Empathy. By the logic behind the metaphor of Morality As Happiness, the inability to experience pleasure and aesthetic insensitivity are moral flaws, since they limit the experience of joy and hence limit one's capacity for empathy and one's ability to give joy to others. Uncommunicativeness and uncooperativeness greatly limit one's capacity to nurture social ties. Lack of self-respect makes it difficult to develop one's full potential, which in turn may keep one from developing one's full capacity for nurturance.

Inquisitiveness and honesty jointly characterize the passion for truth and knowledge—even truth and knowledge about ourselves and our society that may not be pleasant. The ability to nurture successfully requires that we know and understand ourselves and our society—especially the dark side—as deeply and truthfully as we can. Art is partly a matter of beauty—the creation of it and the inquiry into what it is—but just as importantly a matter of inquiring into, exploring, and attempting to comprehend our souls and our society—the dark as well as the light. From the perspective of nurturance, art and the search for knowledge and understanding are, for these reasons, moral activities of the highest order. From the perspective of nurturance, the age-old equation of the Good, the True, and the Beautiful makes perfect sense.

Many of the sins—the moral weaknesses—of Strict Father morality are not present in Nurturant Parent morality. Given the view of Morality As Happiness, the pleasures of the body take on positive moral value, so long as they don't interfere with nurturance, self-nurturance, and the development of one's potential. Sensuality is a virtue, just as is aesthetic sensitivity. Neither are virtues in Strict Father morality. Sex education is important in Nurturant Parent morality, not just to prevent unwanted pregnancy or the transmission of sexual diseases, but also to spread knowledge about nurturant sexuality and how to maximize the giving and receiving of sexual pleasure. Sexual activity without marriage is not immoral in itself; it is immoral only if it results in harm to oneself or others.

MORAL SELF-INTEREST

The metaphor of Moral Self-Interest plays a significant role in Nurturant Parent morality, but it is subservient to and limited by the other metaphors in the system. Moreover, its application is often misunderstood. First, Moral Self-Interest is often confused with Moral Self-Nurturance, which is taking care of yourself so that you can take care of others. Second, it is commonly confused with Morality As Happiness. The idea of being as happy as you can so that you can properly empathize with, and thus be nurturant toward, others is a very different idea than just seeking Self-Interest, especially seeking wealth and power for their own sake. Third, the idea of maximally developing your potential so as to be able to be nurturant toward others is again a very different idea than just seeking Self-Interest.

The difference can be seen in the example of becoming a doctor so as to best serve your community versus becoming a doctor just to get rich. A doctor who serves her community may happen to get rich, and a doctor who gets rich may happen to serve her community, but the difference in the nature of one's relation to one's community and to one's own sense of morality is all-important.

To someone who does not comprehend Nurturant Parent morality, Morality As Self-Nurturance, Happiness, and Self-Development might be confused with Moral Self-Interest. Indeed, liberal theorists often fail to make the distinction, since they correctly see self-nurturance, happiness, and self-development as requiring protection against interference by the state. What they incorrectly infer is that self-nurturance, happiness, and self-development concern the independence and autonomy of the individual, which would make them fall under self-interest. But what is involved in such concepts is interdependence, not independence. Within a community, Morality As Nurturance requires an interdependence among community members. Similarly, all the forms of morality supporting nurturance are in the service of that interdependence, including self-nurturance, happiness, and self-development. The views of liberal theorists that these are forms of Moral Self-Interest, and hence of autonomy and independence, is not true within the system of Nurturant Parent morality.

With all these constraints on the application of Moral Self-Interest within Nurturant Parent morality, there is still plenty of room for its application. If one is serving the cause of nurturance in all these ways, it is fine to seek one's self-interest.

NURTURANCE AND BUSINESS

The relationship between nurturance and moral self-interest can be seen most clearly in nurturant forms of business practice. It involves the humane treatment of employees, the creation of a safe and humane workplace, social and ecological responsibility, fairness in hiring and promotion, the building of a work community, the development of excellent communication between employees and management and between the company and its customers, opportunities for employee self-development, a positive role in the larger community, scrupulous honesty, a regard for one's customers and for the public, and excellent customer service. Policies such as these have increased the productivity and success of many businesses. They are models of how Nurturant Parent morality can function to help businesses be successful and to allow owners, investors, and employees to seek their self-interest within this moral system.

Moral Self-Interest does function in Nurturant Parent morality, but it has a very different meaning, all the forms of nurturance combining to shape what counts as self-interest, especially in a business context.

NURTURANCE AND WORK

Let us understand from the beginning that nurturance *is* work, hard work—that is, the nurturance of children in a family. Within the context of business and of earning a living, Nurturant Parent morality has something very different to say about work than does Strict Father morality, where work is the application of self-discipline for the sake of self-reliance. In that morality, whatever the work is like, it is moral in itself; and if work imposes hardship, well, hardship is good for you, since it builds character.

But Nurturant Parent morality says something different about what work ought to be and the kinds of jobs that ought to exist in a nurturant society. First, Morality As Self-Nurturance says that working at an unsafe or unhealthy job is not necessarily moral. Hence, work should be as safe and healthy as possible, and worker safety should be a major priority. Second, Morality As Self-Development says that work should promote and not impede personal development;

thus, employers, whenever possible, should have such things as educational programs and other personal development programs or should try to arrange employment to allow workers to take part in such programs. Third, Morality As Nurturance implies that work should maximally promote family life and stable communities through, say, parental leave policies, day-care centers, flexible hours, and not forcing employees to relocate over and over again. It also implies that work should maximally protect and enhance the environment. Work that pollutes rivers, destroys rainforests, depletes the ocean's fishing stocks, and so on is not moral work. Fourth, Morality As Happiness implies that work should not be alienating, or boring, or deadening to the human soul and to one's aesthetic consciousness. Work should rather be as enjoyable and rewarding in itself as possible. Additionally, workplaces should make aesthetics a consideration in the conditions of work. Fifth, Morality As Empathy says that work should promote empathetic contact with other people as much as possible. It should not just be working at a machine all day. Sixth, Morality As Fairness implies that people should be paid fairly in proportion to their work.

In short, Nurturant Parent morality has many implications for how work should be set up in a society and for what the dignity of work is. Supplying jobs of some kind or other is not enough. A nurturant society cares about the kinds of jobs they are and what consequences they have—not just what they pay! But it also cares very much what they pay and how equitable pay is for the amount of work done.

NURTURANT MORAL BOUNDARIES

We have just seen that the metaphors of Moral Strength and Moral Self-Interest apply with very different consequences when subordinated to the rest of the Nurturant Parent system. The same is true for other metaphors that have high priority in the Strict Father moral system. Take the metaphor of Moral Boundaries, where action is conceptualized as motion, and motions away from certain paths and out of certain regions is prohibited. The effect of the metaphor is to say that certain specific types of actions are prohibited or required and that violations of those prohibitions or requirements are dangerous to society because they gradually change the prevailing mores in an immoral direction.

There is, not surprisingly, a version of the metaphor of Moral Boundaries in Nurturant Parent morality. The statement of the metaphor is exactly the same as it is in the Strict Father model. But its role in the service of the metaphors of Morality as Empathy, Nurturance, and the rest, changes how it applies. Instead of just specific kinds of actions being strictly prohibited or required, there are in the Nurturant Parent model prohibitions against actions with anti-nurturant consequences. For example, actions that are likely to lead to an impairment of people's health are immoral in Nurturant Parent morality. Such actions are transgressions, pure and simple. Examples include allowing poisonous chemicals to be dumped in public water supplies or inducing teenagers to smoke and thereby develop a cancer-causing tobacco addiction. These are moral transgressions. Such kinds of actions are over the line of moral behavior.

RESTITUTION AND RETRIBUTION

In the Nurturant Parent model of the family, a just way of dealing with a child's moderate violation of proscribed behavior is restitution, not retribution—to have the child help perform some helpful or otherwise nurturant act. But Nurturant Parents also are responsible for protecting their children, and they protect their children fiercely. They want retribution against people who hurt children—against polluters, drug dealers, manufacturers of products known to be unsafe, and so on. Thus, in the cases of harm to children they advocate retribution, but in the cases of impermissible acts by children, they favor restitution. Thus, they too use moral accounting to characterize justice, but the details are different.

THE MORAL AUTHORITY OF THE NURTURANT

Nurturing parents within the Nurturant Parent family have and deserve parental authority. Nurturance is a precondition to authority, and indeed is seen as productive of authority: a fully nurturant parent deserves to be listened to. The same is true of leaders who fulfill their nurturant obligations—who are empathetic, who successfully help people, who are fair, who communicate effectively, and who nurture social ties successfully. Such leaders deserve moral authority. But, in a nurturant morality, moral authority is not the ability to set rules and the responsibility for setting them. Rather it has to do with trust, the trust that a leader will communicate effectively, arrange for participation, be honest, and have the wisdom, experience, and strength to succeed in helping.

NURTURANCE AND EVOLUTION

Evolution is sometimes mistakenly seen in terms of survival of the fittest. Such a view ignores nurturance. No species survives if it does not successfully nurture its young. Evolution can be thought of in terms of nurturance—the survival of species that adapt so as to continue successful nurturance. This idea may not change the theory of evolution itself, but it does change its metaphorical applications.

I mention this because the idea of evolution is sometimes imbued with Strict Father morality: the species that survive are the fittest and strongest, the ones most successful in pursuing self-interest. This Strict Father interpretation of evolu-

tion can then be turned metaphorically into Social Darwin-
ism, the survival of the fittest in society; and then, via the
metaphor of the Moral Order Is the Natural Order, the social
survival of the fittest can be seen as moral. None of this
makes any sense under the interpretation of evolution in
terms of nurturance. If evolution is understood in terms of
the survival of the nurturant, it becomes nonsense to think
of the social survival of the fittest as moral on the grounds
of an evolutionary metaphor for society plus the Moral Order
metaphor.

The Structure of the Model

Like Strict Father morality, Nurturant Parent morality is an
elaborate and important moral system centered around a central, idealized model of the Nurturant Parent family. The
centerpiece of the system is the metaphor of Morality As
Nurturance, which extends morality in family life to morality
in general. This metaphor induces other metaphors by struc-
turing morality in terms of concepts intimately related to
nurturance, such as empathy, self-nurturance, the nurturance
of social ties, self-development, happiness, and fairness.
These metaphors have the highest priority in the system.

MORALITY AS NURTURANCE: This is the most direct expression of the nurturance ethic.

MORALITY AS EMPATHY: Empathy, as a precondition for
nurturance, is of preeminent importance.

MORAL SELF-NURTURANCE: Self-nurturance is a necessary
condition for nurturance.

MORALITY AS THE NURTURANCE OF SOCIAL TIES: This is
necessary for nurturance within a wider community.

MORALITY AS SELF-DEVELOPMENT: Since the development
of the child's potential is a major object of nurturance,
self-development is a major aspect of a nurturant morality.

MORALITY AS HAPPINESS: Because unhappy people are less likely to be empathetic, the cultivation of one's own happiness is crucial to the cultivation of empathy.

MORALITY AS FAIR DISTRIBUTION: Just as nurturance requires fairness of distribution among one's children, so moral nurturance requires this metaphor.

MORAL GROWTH: Since an objective of literal nurturance is physical growth, an objective of moral nurturance becomes moral growth.

MORAL STRENGTH: Strength is critical to nurturance. In childrearing, strength serves nurturance; in morality, moral strength is in the service of morality conceptualized as nurturance. The subservient role of moral strength in this system vastly affects its meaning.

RETRIBUTION AND RESTITUTION: Nurturance requires protection, and fiercely nurturant parents seek retribution against those who would harm children. But when children transgress, nurturance requires the preference of restitution over retribution.

MORAL BOUNDARIES: Nurturant morality produces a different form of transgression.

MORAL AUTHORITY: It arises from your track record as a nurturer.

These metaphors for morality, structured in this way, jointly entail a major mode of moral thought. Here is what each contributes.

MORALITY AS NURTURANCE: From this, it follows that helping people in need of help is moral.

MORALITY AS EMPATHY: Empathy is a projection of your capacity to feel onto another person. An empathetic person will therefore not want others to experience a lack of well-being. And a truly empathetic person will be able to feel what it is like to have another person's values and to see the world from their perspective. According to this metaphor, this is a moral activity and should lead one to be nurturant.

MORAL SELF-NURTURANCE: This states that it is moral to take care of oneself; otherwise, one cannot nurture others and instead imposes obligations on others.

MORALITY AS SELF-DEVELOPMENT: This makes the development of human potential, in oneself and others, into a moral calling.

MORALITY AS HAPPINESS: This creates an anti-ascetic morality and turns the capacity for aesthetic experience into a virtue. Since communing with the natural world is a major aesthetic experience, this makes nurturance toward nature a form of morality.

MORALITY AS FAIR DISTRIBUTION: This brings issues of equality and equitability into the moral system.

MORAL GROWTH: Nurturance promotes moral growth, and although moral growth doesn't always occur if one is too morally stunted, nonetheless it is taken to be possible in a wide range of cases.

MORAL STRENGTH: This stresses protection and, in the context of the system, creates a host of virtues and moral failings. The moral failings are: lack of social responsibility, selfishness, self-righteousness, narrow-mindedness, inability to experience pleasure, aesthetic insensitivity, lack of curiosity, uncommunicativeness, dishonesty, insensitivity to feelings, inconsiderateness, uncooperativeness, meanness, self-centeredness, and lack of self-respect. The virtues are the opposites: social responsibility, generosity, respect for the values of others, open-mindedness, a capacity for pleasure, aesthetic sensitivity, inquisitiveness, ability to communicate, honesty, sensitivity to feelings, considerateness, cooperativeness, kindness, community-mindedness, and self-respect. A person of good character is a person who has these virtues.

MORAL BOUNDARIES: These are defined by actions that produce nonnurturant effects.

MORAL SELF-INTEREST: This is redefined by its subservient
role within this system. The violation of nurturant ethics is
not in anyone's self-interest in this system. Morality As
Self-Nurturance, Morality As Happiness and Morality As
Self-Development take priority over Moral Self-Interest
and are *not* instances of it. Moral Self-Interest, as constrained by nurturant morality, characterizes a morality of
business.

MORAL AUTHORITY: Moral authority accrues by virtue of
successful nurturance and the responsibility for nurturance.
It is not the ability to set and enforce rules; it is earned
trust.

The priorities that we saw in the Strict Father model are
reversed here. Suppose we use the term "The Nurturance
Group" for Moral Nurturance, Moral Empathy, the Nurturance of Social Ties, Moral Self-Development, Moral Happiness, and Morality as Fair Distribution. The hierarchy of
moral values in Nurturant Parent morality can be expressed
as:

> The Nurturance Group
> Moral Self-Interest
> The Strength Group

This is just the reversal of the priorities we found in the
Strict Father model. In one case, however, there is more
than just a reversal of priorities. Morality As Fairness is in
the Nurturance Group and Moral Order is in the Strength
Group. The priority given to Fairness overwhelms Moral
Order. There is virtually no place left for it to apply, except
in the religious instances where God has moral authority over
human beings. Among human beings, it disappears.

Parameters of Variation

Given the absence of Moral Order, there are only three parameters of variation that apply to the Nurturant Parent model:

1. Linear Scales
2. Moral Focus
3. The Pragmatic-Idealistic Dimension

However, they still provide for a great many variations.

LINEAR SCALES

Almost everything in the model is a matter of degree: empathy, nurturance, self-nurturance, protection, developing one's potential, and so on. As a result there are variants in which certain aspects of the model are overdone or underdone. Too much nurturance is smothering, too little is neglectful. Too little self-nurturance is self-sacrificing and imposing a burden on others; too much may take time and energy from needed nurturance.

MORAL FOCUS

Moral focus interacts with linear scales. Thus a parent whose moral focus is protection will put more energy into protection than into other things. This may result in being overprotective. A parent who puts his moral focus on developing his own potential may put an extreme amount of his energy into that, put little energy into anything else, and become self-centered, irresponsible, and neglectful. A parent whose moral focus is happiness may put most of his energy into being happy and become self-indulgent, irresponsible, and neglectful.

But moral focus does not always result in such "pathological'' variations. It is possible to have protection, say, as a moral focus without being overprotective, or to have self-development as a moral focus without being neglectful. For these reasons, proper "balance" is a constant concern for a nurturant parent. Perhaps the most prevalent metaphor I have encountered among nurturant parents trying to maintain such a balance is that of a "juggler" trying to keep many pins in the air at once.

THE PRAGMATIC-IDEALISTIC DIMENSION

In the central model discussed above, nurturance takes priority over the pursuit of self-interest. That is, nurturance is the end and the pursuit of self-interest is a means to serve the goal of being properly nurturant. That is the idealistic version of the model. This is true of parents, who may seek money and power to be able to take care of their family better. It is also true of the raising of children: children are raised to become nurturant, and they learn to pursue their self-interest so that they can be better nurturers, that is, to be able to support their family, develop their own potential, help their children develop their potential, to take care of themselves, and so on.

The pragmatic version reverses ends and means. In the pragmatic version, the pursuit of self-interest is the end and nurturance is the means. You can pursue your self-interest better if you are empathetic, take care of others, take care of yourself, develop your potential, protect others, and treat others fairly. In the pragmatic version of the Nurturant Parent model, you nurture your children so that they can pursue their self-interest.

When we apply these models to politics, we will see that all these variations have political correlates.

Part Three

From Family-Based Morality to Politics

7

Why We Need a New Understanding
of American Politics

The Failure of Liberals to Comprehend Conservatism

We are a few steps away from our denouement, from showing in detail how such an analysis of family-based moral
systems contributes to an answer of the puzzles we started
with and sheds light on why conservatives and liberals have
the political policies they have. But first, it would be useful
to show why such an account is needed. Existing attempts
by liberals to understand conservative politics have failed.
We will begin with three analytic failures by liberals:

1. Conservatism is "the ethos of selfishness."
2. Conservatives just believe in less government.
3. Conservatism is no more than a conspiracy of the ultrarich
to protect their money and power and to make themselves
even richer and more powerful.

THE SELFISHNESS HYPOTHESIS

Let us begin with the mistake of Michael Lerner of *Tikkun*
magazine, whose "politics of meaning" has been endorsed

by Hillary Rodham Clinton. Lerner (*Tikkun,* November/ December 1994, pp. 12, 18) gets some things right: he correctly perceives progressive-liberal politics as being centered on nurturance and community, what he calls "the ethos of caring." But he is quite mistaken when he dismisses conservative politics as being no more than "the ethos of selfishness." He has missed the conservative moral vision and missed the fact that American voters appear to be responding to that moral vision.

If Lerner were right, simple pragmatic appeals to self-interest should work on conservatives. They don't. If he were right, conservatives in California would have endorsed the Single Payer Initiative, since it would have saved them money. If he were right, conservatives would not be endorsing the replacement of AFDC welfare payments with orphanages, since orphanages cost more than AFDC does. If he were right, conservatives would not be endorsing the Three Strikes legislation and all the money to be spent on prisons that it entails. Simply pointing out to conservatives that these policies do not serve their selfish interests should end the matter right there. It has been pointed out, to no effect.

Lerner's "ethos of selfishness" hypothesis does not explain the moral fervor of the conservative majority as it took over Congress at the beginning of 1995. It does not explain the focus on family values. It doesn't even explain why the conservatives advocate the death penalty, or why they want to abolish the NEA, or why they oppose abortion. The selfishness hypothesis simply does not explain conservative policies.

THE LESS-GOVERNMENT HYPOTHESIS

Why does conservative politics take the shape it does? Why should conservatives be proposing orphanages? Abolishing the Environmental Protection Agency? Abolishing the arts and humanities endowments? Is it merely, as is repeated over and over, that conservatives want less government and liberals want more?

That cannot be true. Conservatives don't merely want less government. They want to raise spending for the military—even bring back Star Wars—not reduce it. They want to build more prisons. There is no move to eliminate the drug enforcement agency. Or the FBI, or the intelligence agencies. There is no outcry to stop bailouts of large corporations, like Lockheed. Or eliminate nuclear power development. Or to stop funding computer research. There is no attempt to charge airlines for the training of pilots by the Air Force. Or to charge automobile companies for the building of highways. If conservatives simply wanted less government spending or wanted government to pay for itself, there are a myriad of other cuts and reforms they could be proposing. The Less-Government Hypothesis is simply false. It does not explain what conservatives do and don't want to spend money on. Conservatives want to spend on some things and not others. What determines which ones?

THE CYNICAL LIBERAL RESPONSE

Anthony Lewis (*New York Times* op-ed page, February 27, 1995) lists the following conservative budget cuts: repeal of the National School Lunch Act; ending the WIC (Women, Infants, and Children) program that has reduced infant mortality by providing nutrition to impoverished mothers and children; and legislation making it harder for investors to sue in cases of securities fraud. He comments:

> Looking at that list of actions taken and planned, one can hardly miss the theme. The purpose of one measure after another is to enrich those who have money and power in our society and reduce the modest help

this country gives to the poor and the weak. Manufacturers and drug companies would gain. Sick children and poor mothers would lose.

This is an example of the cynical liberal response to conservative government.

The cynical liberal response is that conservative politicians are all tools of the ultrarich and the big, multinational corporations the rich control. Under the Reagan and Bush administrations, there was a massive redistribution of wealth toward the ultrarich, so that now the top 10 percent of families control 70 percent of the nation's wealth. The Reagan administration added three trillion dollars to the national debt, and redistributed it to the ultrarich, making the rest of the country pay interest on the debt, which amounts to 28 percent of the federal budget every year.

The cynical liberal response is that conservatives want to continue spending on (1) the means of social control such as the military, the police, the intelligence services, and prisons, and on (2) aspects of government that help make the rich richer, say, the funding of computer research, or nuclear power, or the Air Force's training of pilots which benefits the airlines, or the bailouts of large corporations.

The cynical liberal response is that the ultrarich are attempting to take over the intellectual life of the country to ensure their domination. One step has been to finance a network of right-wing think tanks. Eliminating the National Endowment for the Humanities would eliminate a major source of funding for non-right-wing research. Eliminating the Corporation for Public Broadcasting would curtail public discourse in a way that would serve thought control. Controlling the purse strings of public universities would be another step in thought control. Setting the agenda for moral education would be still another.

There is much to be said for the cynical liberal response.

Much of it is true. Yet it has major flaws and is far from the whole story. First, it is a demonization of conservatives. It assumes that they are either rich, evil, self-serving power-mongers, or their paid agents, or dupes. The conservative ranks may well contain some of each. Yet most conservatives are not rich and see themselves as working for the benefit of the country rather than for their own benefit. There are too many idealistic conservatives of good intentions and moderate means for the demonization theory to be true.

Second, the conspiracy theory attributes too much to competence and to centralized control. Political life in America is not run from the top by a smooth-functioning machine. It is messy. American politics is not something that yields readily to rational control. A well-financed smooth-functioning machine can do a lot in political organization and propaganda, but it cannot implant a totally different worldview in tens of millions of minds. It must use ideas that are already there and well respected in the culture.

Third, the conspiracy theory does not explain why conservative rhetoric can make sense to so many people who did not previously vote conservative. It does not explain why such people simply did not experience cognitive dissonance and disbelief when they heard the campaign rhetoric. The cynical liberal explanation is the Orwellian one, that any Big Lie repeated often enough will be believed. But that assumes an old-fashioned stimulus-response view of the human mind that both ignores what is known about the human brain and ignores the effects of culture. We are all immersed in American culture. Our cultural knowledge is physically encoded in the synapses of our brains. People do not get new worldviews overnight. New ideas are never entirely new. They must make use of ideas already present in the culture. No conspiracy of the ultrarich explains why conservative ideas make sense to people and what sense they make.

Fourth, the conspiracy theory does not explain the details

of conservative political positions. Why should the death penalty be in the interest of the ultrarich? How can the rich get richer on the Three Strikes and You're Out law, which requires heavy government spending on prisons? How would orphanages serve the interests of the ultrarich? Why should the ultrarich want to get rid of the National Endowment for the Arts? The conspiracy theory simply doesn't explain many important conservative policies.

Moreover, even where the ultrarich do benefit from conservative policies, a deeper explanation is in order. Why should conservative morality serve ultrarich interests? What links are there between conservative family values and the interests of the ultrarich? Simply positing a conspiracy of the ultrarich does not answer these questions.

In short, I do not believe the cynical liberal claim that the details of conservative political policies are just due to a self-serving ultrarich conspiracy, though the interests and finances of the ultrarich are certainly engaged. Indeed, I have not heard any liberal account of conservatism that makes sense of conservative policies, or the conservative worldview, or conservative language. I think there is a deeper explanation that comes out of the cultural role of the Strict Father model of the family and the moral schemes that fit that model.

The Conservative Failure to Understand Conservatism

Even the views of conservative thinkers don't really help in characterizing what conservatism is. There are three principal conservative descriptions of conservatism.

1. Conservatism is against big government.
2. Conservatism is for traditional values.
3. Conservatism is just what the Bible tells us.

We have already seen that the first is false. As for the second, take what William J. Bennett, one of the major conservative intellectuals, says:

> Conservatism as I understand it . . . seeks to conserve the best elements of the past. It understands the important role that traditions, institutions, habits and authority have in our social life together, and recognizes our national institutions as products of principles developed over time by custom, the lessons of experience, and consensus. . . . Conservatism, too, is based on the belief that the social order rests upon a moral base. (References, C1: Bennett 1992, p. 35)

Bennett's account doesn't help much. It doesn't say what is to count as the "best" elements of the past and why. Racism, colonialism, witch-burning, child labor, and even the sale of children as indentured servants are not among the "best" elements of American tradition. But it is not clear by what criterion something is to count as "best." Bennett mentions traditional institutions, but government and public schools are not traditional institutions that count for conservatives. He mentions consensus, but conservatives support views where there is no consensus—anti-abortion legislation, the abolishment of social programs, and so on. He mentions a "moral base" but gives no general account of why conservative views of morality are to count as "moral," while liberal views of morality are not to count as "moral."

The same problem inheres in the claim of right-wing religious groups who state that conservatism is just a matter of following the Bible. The Bible cannot be applied to politics or much else without a lot of selection and interpretation. The National Council of Churches also urges following the Bible, but gives it a liberal interpretation. Liberation theology also follows the Bible, with an often revolutionary inter-

pretation. What, exactly, characterizes a conservative interpretation of the Bible? Until this prior question is answered
adequately, it will be hard to understand just which Christians see their religion as fitting conservative politics and
why. We will discuss this in Chapter 14.

What all this suggests is that conservatives themselves are
not particularly good at characterizing what unifies their own
political philosophy. Nor does it appear that liberals are any
better at characterizing political liberalism. Theoreticians of
liberalism see their job as normative, not descriptive, as saying what liberalism should be rather than describing what it
actually is. Not surprisingly, the normative theoretical characterizations of liberalism do not do a very good descriptive
job. Thomas Spragens, Jr., provides a typical view:

> The essence of liberalism as a normative doctrine is
> its focus on the protection of rights as the central (per-
> haps the only) purpose of political society. Its essence
> as a social theory is its focus on autonomous and sepa-
> rate individuals as the sum and substance of society.
> A properly ordered society, therefore, is centered
> around contractural relationships among these individu-
> als. (References, C4: Spragens 1995)

This does not in any way distinguish between contemporary
liberals and conservatives. The question to be asked is
"Which rights count?" Conservatives declare the right to
keep what you've earned, the right to own machine guns,
the rights of the unborn, the right to do anything you want
with your property, the right to form a private heavily armed
militia, and so on. If it is liberals who fear the coercive
power of the state, why is it that conservatives are trying to
destroy federal power and liberals are trying to preserve it?
Without an account of what rights count and what coercive
powers of the state are bad, the classical theory of liberalism
cannot distinguish political liberalism from conservatism.

Other classic liberal theories focus upon liberty and equality jointly. Rawls, for example, adds to liberty an account of equality in which any inequalities must benefit the most disadvantaged members of society. This tells us nothing about why political liberals favor ecology, why they are not anti-abortion, why they defend funding for the arts, and so on. From the abstract realms of liberty and equality, you can't get down to the nitty-gritty of real political stands on issues.

The communitarian critiques, on the whole, don't do much better than the classical liberal views. They correctly point out that the classical liberal myth of the autonomous individual entering into social contracts with other autonomous individuals doesn't make much sense. Individuals are not and never were autonomous. We are social through and through, and social life necessarily demands responsibilities as well as rights. But which responsibilities and why? Conservatives also stress responsibility. What's the difference?

Another common claim has to do with the liberal and conservative views of human nature: conservatives think that people are basically rotten and have to be subject to authority and disciplined, while liberals think that people are basically good and can decide what to do for themselves. That theory just doesn't jibe at all with contemporary liberal and conservative politics. Liberals don't think that people out to maximize their profit can be counted on to do the right thing—not to pollute, not to create unsafe working conditions, not to make unsafe products, not to discriminate unfairly. It's the liberals who are suspicious of human nature on many issues and the conservatives who are trusting.

Michael Lerner, as noted above, is on the right track when he talks about "the ethos of caring" as being central to liberalism. But he does not spell out just what the details of that ethos are and why it leads to the particular stands that liberals tend to hold. Moreover, conservatives, too, "care" about many things—the morals of their children, the rights of the unborn, what is taught in our schools, the victims of crimes, the effects of our society on sex, drugs, and violence. How does the caring of conservatives differ from the caring of liberals? It is not caring alone that makes the difference.

I believe that the answer, or at least a large part of it, has to do with Strict Father and Nurturant Parent morality. I will argue that these opposed moral visions lie behind the worldview differences between conservatives and liberals. I will also argue that variations on these moral systems can explain the rich variety of positions within each camp.

The remaining step in the argument remains to be taken: what links the family and family-based morality to politics?

8

The Nature of the Model

The Nation As Family Metaphor

Part of our conceptual systems, whether we are liberals, conservatives, or neither, is a common metaphorical conception of the Nation As Family, with the government, or head of state representing the government, seen as an older male authority figure, typically a father. We talk about our founding *fathers*. George Washington was called "the father of his country," partly because he was the metaphorical "progenitor" who brought it into being and partly because he was seen as the ultimate legitimate head of state, which according to this metaphor is the head of the family, the father. The U.S. government has long been referred to as "*Uncle* Sam." George Orwell's nightmare head of state in *1984* was called "Big *Brother*." This has been consciously echoed in the conservatives' use of "big government." When our country goes to war, it sends its *sons* (and now its *daughters*) into battle. A *patriot* (from the Latin *pater*, "father") loves his *fatherland*. We ask God in song to "crown thy good [i.e., the good of the nation] with *brotherhood*." The metaphor even comes up in legislative argument. Senator Robert

Dole, in arguing for the balanced-budget amendment, chided liberals as thinking that "Washington knows best," a slogan based on the cliché "Father knows best," which had also been the title of a popular TV show.

Indeed, an argument regularly used for the balanced-budget amendment is that, just as a family's budget must be balanced, so must a nation's. Any economist, liberal or conservative, knows that there are many crucial differences between a family and a nation that make the analogy economically ludicrous: a family can't initiate economic stimulus programs, print new currency, or increase tax rates. Yet, despite this, the unconscious and automatic Nation As Family metaphor in our conceptual systems makes the logic seem to be just commonsense to most people.

My point is that the Nation As Family metaphor exists as part of our standard conceptual repertoire. I believe it does a lot more conceptual work than just allowing us to make sense of expressions like "Uncle Sam" or "Big Brother" or permit advocates of the balanced-budget amendment to get away with conceptualizing the nation as a family in their arguments. I believe that the Nation As Family metaphor is what links conservative and liberal worldviews to the family-based moralities we have been discussing. I believe that this metaphor projects the Strict Father and Nurturant Parent moral systems onto politics to form the conservative and liberal political worldviews.

A BIT MORE PRECISION

It's time to get a bit more precise about the model proposed. First, the Nation As Family metaphor can be stated as follows (here, for simplicity, we limit the older authority figure in the family to a parent):

- The Nation Is a Family.
- The Government Is a Parent.
- The Citizens Are the Children.

This metaphor allows us to reason about the nation on the basis of what we know about a family. For example, just as a parent functions to protect his or her children, so the government functions to protect its citizens. Certain inferences, importantly, are overridden, as is normal in conceptual metaphors. For example, citizens, for the most part, are adults and so are not treated like children. The government doesn't put you to bed, tell you a bedtime story, and so on. This is predicted by what is called the Invariance Principle (References, A1, Lakoff 1993). However, the government, like a parent, does have certain responsibilities toward its citizens and authority over them.

Notice that this metaphor does not specify exactly what kind of family the nation is. This is where the Strict Father and Nurturant Parent models come in; they fill in such information. For conservatives, the nation is conceptualized (implicitly and unconsciously) as a Strict Father family and, for liberals, as a Nurturant Parent family. The link between morality and politics arises as follows: The Strict Father and Nurturant Parent models of the family induce the two moral systems discussed in Chapters 5 and 6. The Nation As Family metaphor, in applying to the family models, also applies to the family-based moral systems, yielding conservative and liberal political worldviews.

Described from the ground up, this analysis of conservative and liberal worldviews may seem elaborate, but from the perspective of the structure of conceptual systems, it is actually very simple. Each of the elements in the analysis exists independently:

1. The two models of the family, which are culturally elaborated variants of traditional male and female models. These are rooted in long cultural experience.
2. The various metaphors for morality, in which morality is conceptualized as strength, nurturance, authority, health, and so on. These are grounded in everyday experiential well-

being: it's better to be strong rather than weak, cared for rather than not cared for, in control rather than not in control, healthy rather than sick, and so on.

3. The Nation As Family metaphor.

These elements, which exist independently, fit together naturally in certain ways. Each of the family models provides a natural organization of the metaphors for morality, as described in Chapters 5 and 6. The result is two opposing moral systems. The Nation As Family metaphor projects these two moral systems onto the domain of politics, yielding the conservative and liberal worldviews. In short, given the independent existence of the two family models, the metaphors for morality, and the Nation As Family metaphor, these two political worldviews are the minimal ways of using these conceptual elements to arrive at an approach to politics. The conservative and liberal worldviews are the results of a maximally economic use of existing conceptual resources to make sense of politics. And as we shall see below, variations on liberal and conservative worldviews are minimal variations on these models. But variations aside for the moment, the two worldviews are each very simply constituted. Each is a binding together of three kinds of independently existing elements. From the perspective of the human brain, this is very simple indeed.

EXPLANATION AND EVIDENCE

The kind of analysis I am presenting is known as cognitive modeling. It is perhaps the most common form of analysis within the cognitive sciences. The idea is to construct a model of how the mind, using natural cognitive apparatus (such as conceptual metaphors and radial categories), makes sense of some significantly wide range of phenomena, especially puzzling phenomena.

Plausible models have the kinds of properties that this model has. The most plausible models are those whose elements have an independent motivation and use minimal additional cognitive apparatus. The plausibility of the model rests on the plausibility of other claims. First, that the idealized models of the family presented are really cognitive stereotypes. Second, that the analysis of the metaphors for morality is a plausible one, based on evidence from inference and language. Thus, do we really understand morality as purity, or strength, or nurturance, and how can we tell? Some of that inferential and linguistic evidence was given above in the discussion of the metaphors. And plausible experiential bases for those metaphors were presented in Chapter 3. Third, is it plausible that our conceptual systems have a metaphor for conceptualizing a nation as a family? The considerations at the beginning of this section do seem to justify that conclusion. As conceptual analyses go in our discipline, this one has a high degree of initial plausibility. That is, it is the kind of model a cognitive linguist would expect to find.

The next question is whether the model accounts for the phenomena. These phenomena were discussed in Chapters 1 and 2. They are of three kinds. First, the model must explain why conservative and liberal political stands group together as they do. Take, for example, opposition to social programs, anti-environmentalism, anti-feminism, harsh penalties for criminals, and support of the right to own assault weapons. Why do they fit together? Second, the model must explain what puzzles liberals about conservatives and conservatives about liberals. It must explain why contradictions for one are obvious truths for the other. Third, it must account for the details of conservative and liberal discourse. It must account for how texts fit together and make sense, and it must account for how metaphorical language is used in those texts. Moreover, the model must be predictive. It must account for the modes of reasoning and metaphorical language in new texts—texts not yet produced. It must account for how conservatives and liberals come down on new issues. And it must account for new puzzles that arise. Getting any cognitive model to do all this is a tall order.

Very few of those outside the cognitive sciences are used to thinking about social and political issues in terms of the human mind. It is common to think of them in terms of economics or sociology or political philosophy or law or statistical studies that use survey data. To date none of those accounts, to my knowledge, have been able to make complete sense of the three kinds of phenomena considered here. So far as I have been able to discover, this hypothesis is the only serious attempt to explain all these phenomena together.

Since this hypothesis is new, it does not have the degree of confirmation that one would expect of more mature theories. At present, it is based solely on modeling—on whether the model is plausible and how well it accounts for the three types of data discussed. It appears to fit extremely well and to have held up predictively so far. Virtually every talk show and political speech I've listened to since working this out has confirmed the predictions of the model. That, to a cognitive modeler, is very strong empirical confirmation. But anyone would prefer to have additional confirmation, from, say, psycholinguistic tests and from survey data, if possible. I hope such studies can be undertaken in the future, but they would not be easy or straightforward. Psycholinguistic testing has begun to be able to discern the existence of conceptual metaphors in cognitive models, but no experimental paradigms of the complexity needed to test this hypothesis now exist (see References, A1, Gibbs 1994). Survey research has not yet developed an adequate methodology to test for the presence of complex metaphorical cognitive models such as these.

Let us now shift from discussing this analysis in terms of

cognitive models, evidence, explanation, and prediction to what it says about people. The analysis claims that we use unconscious cognitive models to comprehend politics, just as we use them in all other areas of our lives. Whenever we instantaneously understand a political speech, we are filling in what is not explicitly said in the speech through the use of these cognitive models. This analysis claims that the difference between conservative and liberal worldviews derives from different cognitive models of politics. The most fundamental difference, the analysis claims, the difference from which all other differences spring, is in the use of an idealized, stereotypical model of the family. The conservative model uses a Strict Father model of the family, while the liberal model uses a Nurturant Parent model of the family. Both conservative and liberal models then organize and prioritize common conceptual metaphors for morality so as to fit the family model. The resulting family-based moralities are linked to politics by a common Nation As Family metaphor. The result is two very different political worldviews.

It is important to note what the analysis does *not* claim. It does not claim that each person has only one idealized family model. Most of us probably recognize both models and use them differently. We may believe one and mock the other (though to mock it, we have to recognize it). Another possibility is that we have both models and use them differently, applying one model to family life and the other to politics.

I would not be surprised if many people applied the Strict Father model to how fathers should act and the Nurturant Parent model to how mothers should act. They may then have a model of the family with both a Strict Father and Nurturant Mother, with a separation of responsibilities and each functioning differently. Where the models contradict each other in family life, as they inevitably do, there are many choices for resolution. Perhaps the father's model takes precedence, perhaps the mother's does, perhaps it is argued out on a case-by-case basis, or perhaps it just depends on who has the most energy that day. Now such a family, with different people using two distinct idealized family models, cannot be the basis for a coherent politics. To arrive at a coherent political worldview via the Nation As Family metaphor, one of the family models must be chosen, in the way the analysis indicates.

Of course, as I mentioned in Chapter 1, people do not necessarily have a single, coherent worldview based on a single model. For example, from 1968 to 1992 (with the exception of the Carter presidency), the voters selected a fairly conservative president and a fairly liberal Congress, creating what might be seen as a Strict Father executive and a Nurturant Mother Congress, thus reproducing a classic family model in the government, with strictness at the top and caring right below.

Thus, the analysis does not claim that there is always, or even mostly, any simple one-to-one correlation between family models and political worldviews. But I suspect that such one-to-one correlations do exist. A conceptual system with such a one-to-one correspondence between family and politics would be simpler, more unified, and more stable (or more rigid), and produce less cognitive dissonance than a system that uses different models at different times on different issues. The conservative focus on family values can be seen from this perspective as an attempt to unify the use of the Strict Father model for family life with its use as a basis for conservative politics. From the perspective of cognitive science, this is an extremely sophisticated and powerful political strategy.

It is time to move on from the general to the particular. Whatever technical or scientific merits this proposal may or may not have, its ultimate value to us as citizens is whether it gives us real insight into our politics, which is what the next chapters are about. I will begin by showing first in Chapter 9 how these two moral systems create different categories of moral actions, of model citizens, and of demons. Then I will move on, in Chapter 10, to answer the questions we started with.

9

Moral Categories in Politics

Categories of Moral Action

A moral system defines how one views the world, how one comprehends hundreds of events, great and small, every day. One of the major ways in which a moral system characterizes worldview is through categorization. Each moral system creates a number of fixed major categories for moral action. Those major categories allow us to classify actions instantly into those that are moral and those that are not, with little or no reflection. Sometimes, we may have trouble fitting an action or event to a category, but mostly we barely notice that we are even classifying. These classifications may sometimes be reflected on consciously and classifications of actions may be changed when we reason consciously. But on the whole, our first unreflective classification stands.

CONSERVATIVE MORAL CATEGORIES

The conservative (Strict Father) and liberal (Nurturant Parent) moral priorities create two different systems for categorizing moral actions. Let us look at them one at a time. Here is the conservative system:

Conservative categories of moral action:
1. Promoting Strict Father morality in general.
2. Promoting self-discipline, responsibility, and self-reliance.
3. Upholding the Morality of Reward and Punishment
   a. Preventing interference with the pursuit of self-interest by self-disciplined, self-reliant people.
   b. Promoting punishment as a means of upholding authority.
   c. Insuring punishment for lack of self-discipline.
4. Protecting moral people from external evils.
5. Upholding the Moral Order.

I have listed five major categories. There may be more, but these are all used a great deal and will suffice for our purposes. Let us look at each category to see where it comes from in the moral system.

1. Promoting Strict Father morality.
   Several metaphors imply a strict good-evil division, in particular, Moral Strength, Moral Boundaries, and Moral Authority. Moral Strength sees evil as a force in the world, reifying it and distinguishing it from good. Moral Boundaries are drawn strictly and clearly between right and wrong. And Moral Authority sets rules to be obeyed, rules that define what is right and distinguish it from what is wrong. The moral system itself, of course, is right—so right that it defines what right is. Defending that system, which defines the very nature of right and wrong, is *the* primary moral obligation. Actions promoting or protecting the moral system are therefore moral; actions against the moral system are therefore immoral.

2. Promoting self-discipline, responsibility, and self-reliance.

The primacy of Moral Strength implies that these are primary virtues. Actions promoting these primary virtues are thus moral; actions discouraging them are therefore immoral.

3. Upholding the Morality of Reward and Punishment

The very notions of reward and punishment are based on the metaphor of moral accounting, as discussed in Chapter 4.

Strict Father morality assumes that it is human nature that people operate in terms of rewards and punishments. Rewards for obedience and punishments for disobedience are crucial to maintaining moral authority; as such, they lie at the heart of this moral system and are thus moral. Actions that uphold the reward-punishment system are therefore moral. Actions against the reward-punishment system are immoral.

There are three important special cases. They are:

3a. Preventing interference with the pursuit of self-interest by self-disciplined, self-reliant people.

The pursuit of self-interest is a system of reward for being self-disciplined and self-reliant, which are primary moral requirements according to Moral Strength. Interfering with this system of reward for being moral is therefore immoral. Preventing such interference is therefore moral.

3b. Promoting punishment as a means of upholding authority.

In Strict Father morality, legitimate authority must be upheld at all costs or the moral system ceases to function. Punishment for violating authority is the main way in which authority is maintained. It is therefore moral to promote punishment for violations of legitimate authority and immoral to act against it.

3c. Insuring punishment for lack of self-discipline.

Moral Strength makes self-discipline a primary moral requirement and the lack of it immoral. Therefore, actions ensuring punishment for moral weakness are moral; actions going against punishment for moral weakness are immoral.

4. Protecting moral people from external evils.

Since protection from external evils is a fundamental part of Strict Father morality, protective actions are moral and inhibiting them is immoral.

5. Upholding the Moral Order.

Since the Moral Order defines legitimate authority, actions upholding it are moral, and actions going against it are immoral.

These categories of moral action greatly facilitate using the moral system. They provide a simplified, conventional way of putting the moral system into practice.

LIBERAL MORAL ACTION

Liberals, too, have categories of moral action, and not surprisingly, they look very different from conservative categories.

Liberal categories of moral action:
1. Empathetic behavior and promoting fairness.
2. Helping those who cannot help themselves.
3. Protecting those who cannot protect themselves.
4. Promoting fulfillment in life.
5. Nurturing and strengthening oneself in order to do the above.

Again, let us look at where they come from, one by one.

1. Empathetic behavior and promoting fairness.

The primacy of Morality as Empathy makes empathy a

moral priority. Morality as Fairness is a consequence; if you empathize with others, you will want them to be treated fairly. This makes empathetic actions and actions promoting fairness into moral actions. Correspondingly, a lack of empathetic behavior, or actions going against fairness, are immoral.

2. Helping those who cannot help themselves.

The priority given to Morality as Nurturance makes it moral to help someone who cannot help himself, and immoral not to do so if one can.

3. Protecting those who cannot protect themselves.

The priority of protection in Nurturant Parent morality makes it moral to protect those who cannot protect themselves, and immoral not to do so when one can.

4. Promoting fulfillment in life.

Moral Happiness and Moral Self-Development make it moral to promote fulfillment in life and immoral to work against it. Fulfillment includes developing your potential in a variety of areas, having meaningful work, being basically happy, and so on.

5. Nurturing and strengthening oneself in order to do the above.

To nurture properly, one must be strong and healthy and must feel nurtured oneself. Therefore acts of taking care of oneself or helping others take care of themselves are moral acts. Since being neglectful of one's health and strength imposes an unfair burden on others, it is immoral not to take care of oneself or to impede others from doing so.

For the sake of comparison, let us look at both moral category systems together:

Conservative categories of moral action:
1. Promoting Strict Father morality in general.
2. Promoting self-discipline, responsibility, and self-reliance.

3. Upholding the Morality of Reward and Punishment.
   a. Preventing interference with the pursuit of self-interest by self-disciplined, self-reliant people.
   b. Promoting punishment as a means of upholding authority.
   c. Ensuring punishment for lack of self-discipline.
4. Protecting moral people from external evils.
5. Upholding the Moral Order.

Liberal categories of moral action:
1. Empathetic behavior and promoting fairness.
2. Helping those who cannot help themselves.
3. Protecting those who cannot protect themselves.
4. Promoting fulfillment in life.
5. Nurturing and strengthening oneself in order to do the above.

These categories define the first moral questions one "asks" unconsciously and automatically of any action. If it is in one of the categories, it is moral; if it is in the opposite category, it isn't moral. Whatever other system of categories one may have—and any conceptual system has a great many—when one is functioning politically, these moral categories are primary. The categories define opposing moral worldviews, worldviews so different that virtually every aspect of public policy looks radically different through these lenses.

Take a simple example: college loans. The federal government has had a program to provide low-interest loans to college students. The students don't have to start paying off the loans while they are still in college and the loans are interest-free during the college years. The liberal rationale for the program is this: College is expensive and a great many poor-to-middle-class students cannot afford it. This loan program allows a great many students to go to college who otherwise wouldn't. Going to college allows one to get a better job at a higher salary afterward and to be paid more during one's entire life. This benefits not only the student but also the government, since the student will be paying more taxes over his lifetime because of his better job.

From the liberal moral perspective, this is a highly moral program. It helps those who cannot help themselves (Category 2). It promotes fulfillment in life in two ways, since education is fulfilling in itself and it permits people to get more fulfilling jobs (Category 4). It strengthens the nation, since it produces a better-educated citizenry and ultimately brings in more tax money (Category 5); and it is empathetic behavior (Category 1) making access to college more fairly distributed (Category 1).

But through conservative spectacles, this is an immoral program. Since students depend on the loans, the program supports dependence on the government rather than self-reliance (Category 2). Since not everyone has access to such loans, the program introduces competitive unfairness, thus interfering with the free market in loans and hence with the fair pursuit of self-interest (Category 3a). Since the program takes money earned by one group and, through taxation, gives it to another group, it is unfair and penalizes the pursuit of self-interest by taking money from someone who has earned it and giving it to someone who hasn't (Category 3a).

I started with college loans because it is not as heated an issue as abortion or welfare or the death penalty or gun control. Yet it is a nitty-gritty issue, because it affects a lot of people very directly. To a liberal, it is obviously the right thing to do. And to a conservative, it is obviously the wrong thing to do. The metaphors for morality that give rise to these inferences are the following. For the liberals: Moral Empathy, Moral Nurturance, Moral Self-Development, and Moral Self-Nurturance. For the conservatives: Moral Strength, Moral Self-Interest, and Moral Accounting (the metaphorical basis of the concepts of Reward and Punish-ment).

The point of this example is that policy debates are not matters of rational discussion on the basis of literal and ob-jective categories. The categories that shape the debate are moral categories; those categories are defined in terms of different family-based conceptions of morality, which give priority to different metaphors for morality. The debate is not a matter of objective, means-end rationality or cost-benefit analysis or effective public policy. It is not just a debate about the particular issue, namely, college loans. The debate is about the right form of morality, and that in turn comes down to the question of the right model of the family. The role of morality and the family is inescapable, even if you are only talking about college loans policy.

Model Citizens and Demons

Conservative and liberal categories for moral action create for each moral system a notion of a model citizen—an ideal prototype—a citizen who best exemplifies forms of moral action.

CONSERVATIVE MODEL CITIZENS

In the conservative moral worldview, the model citizens are those who best fit all the conservative categories for moral action. They are those (1) who have conservative values and act to support them; (2) who are self-disciplined and self-reliant; (3) who uphold the morality of reward and pun-ishment; (4) who work to protect moral citizens; and (5) who act in support of the moral order. Those who best fit all these categories are successful, wealthy, law-abiding conservative businessmen who support a strong military and a strict crimi-nal justice system, who are against government regulation, and who are against affirmative action. They are the model citizens. They are the people whom all Americans should emulate and from whom we have nothing to fear. They deserve to be rewarded and respected.

These model citizens fit an elaborate mythology. They have succeeded through hard work, have earned whatever they have through their own self-discipline, and deserve to keep what they have earned. Through their success and wealth they create jobs, which they "give" to other citizens. Simply by investing their money to maximize their earnings, they become philanthropists who "give" jobs to others and thereby "create wealth" for others. Part of the myth is that these model citizens have been given nothing by the government and have made it on their own. The American Dream is that any honest, self-disciplined, hard-working person can do the same. These model citizens are seen by conservatives as the Ideal Americans in the American Dream.

CONSERVATIVE DEMONS

Correspondingly, conservatives have a demonology. Conservative moral categories produce a categorization of citizens-from-hell: anti-ideal prototypes. These nightmare citizens are those who, by their very nature, violate one or more of the conservative moral categories; and the more categories they violate, the more demonic they are.

CATEGORY 1 DEMONS: Those who are against conservative values (e.g., Strict Father morality). Feminists, gays, and other "deviants" are at the top of the list, since they condemn the very nature of the Strict Father family. Others are the advocates of multiculturalism, who reject the primacy of the Strict Father; postmodern humanists, who deny the existence of any absolute values; egalitarians, who are against moral authority, the moral order, and any other kind of hierarchy.

CATEGORY 2 DEMONS: Those whose lack of self-discipline has led to a lack of self-reliance. Unwed mothers on welfare are high on the list, since their lack of sexual self-control has led to their dependence on the state. Others are unemployed drug users, whose drug habit has led to their being unable to support themselves; able-bodied people on welfare—they can work and they aren't working, so (in this land of opportunity) they are assumed to be lazy and dependent on others.

CATEGORY 3 DEMONS: Protecters of the "public good." Included here are environmentalists, consumer advocates, advocates of affirmative action, and advocates of government-supported universal health care who want the government to interfere with the pursuit of self-interest and thus constrain the business activities of the conservatives' model citizens.

CATEGORY 4 DEMONS: Those who oppose the ways that the military and criminal justice systems have operated. They include antiwar protesters, advocates of prisoners' rights, opponents of police brutality, and so on. Gun control advocates are high on this list, since they would take guns away from those who need them to protect themselves and their families both from criminals and from possible government tyranny. Abortion doctors may be the worst, since they directly kill the most innocent people of all, the unborn.

CATEGORY 5 DEMONS: Advocates of equal rights for women, gays, nonwhites, and ethnic Americans. They work to upset the moral order.

The demon-of-all-demons for conservatives is, not surprisingly, Hillary Clinton! She's an uppity woman (Category 5, opposing the moral order), a former antiwar activist who is pro-choice (Category 4), a protector of the "public good" (Category 3), someone who gained her influence not on her own but through her husband (Category 2), and a supporter of multiculturalism (Category 1). It would be hard for the conservatives to invent a better demon-of-all-demons.

These categories are extremely stable and they resist efforts at change. Secretary of Labor Robert Reich found this out shortly after the 1994 elections, when he sought to recategorize the best model citizens of all—large successful corporations and the people who run them. Reich attempted to use the conservative demonization of welfare recipients against the conservative conception of model citizens. He attacked big corporations and the ultrarich for being recipients of "corporate welfare." Reich pointed out that large corporations owned by the ultrarich receive from the government huge amounts of money that they do not earn: money from inordinately cheap grazing rights, mineral and timber rights, infrastructure development that supports their businesses, agricultural price supports, and hundreds of other kinds of enormous government largesse that come out of the taxpayer's pocketbook—an amount far exceeding the cost of social programs. If the government eliminated corporate welfare, Reich argued, then it could easily afford social programs to help the poor.

Reich's attempt to turn the conservatives' model citizens into conservative demons was doomed to failure, and it fell flat immediately. The reason is clear. The status of successful corporations and the ultrarich as model citizens has become conventionalized—fixed in the conservative mind. They are icons, standard examples to conservatives of what model citizens are. Moreover, they do not fit the stereotype of welfare recipients. They are seen as self-disciplined, energetic, competent, and resourceful rather than self-indulgent, lazy, unskilled, and hapless.

Reich's attempt to call attention to the enormous unearned

largesse bestowed by the government on big corporations failed because he did not really understand the conservative worldview and the cognitive structure underlying American politics. The conservative heroes and demons are what they are for the deepest of reasons, because conservatism rests on a widespread, deeply entrenched family-based moral system. You don't change that with a single speech.

LIBERAL MODEL CITIZENS

Liberals have a very different notion of a model citizen, again generated by liberal moral categories. The ideal liberal citizen is socially responsible, and fits as many of the liberal moral categories as possible. The model liberal citizen (1) is empathetic; (2) helps the disadvantaged; (3) protects those who need protection; (4) promotes and exemplifies fulfillment in life; and (5) takes care of himself so he can do all this. Model liberal citizens are those who live a socially responsible life: they include socially responsible professionals; environmental, consumer, and minority rights advocates; union organizers among impoverished and badly treated workers; doctors and social workers who devote their lives to helping the poor and the elderly; peace advocates, educators, artists, and those in the healing professions. Interestingly, there does not seem to be any identifiable type in American life that is a model citizen in all of these ways. There have certainly been individuals who have been models in one or another of these ways, e.g., Martin Luther King, Jr., Franklin and Eleanor Roosevelt, John and Robert Kennedy, and, for many, Hillary Clinton.

LIBERAL DEMONS

There is, of course, as rich a liberal demonology as there is a conservative one. Those who violate categories 1 to 5 are the monsters of society.

CATEGORY 1 DEMONS: The mean-spirited, selfish, and un-fair—those who have no empathy and show no sense of social responsibility. Wealthy companies and busi-nessmen who only care about profit are at the top of the list, because of their power and political in-fluence.

CATEGORY 2 DEMONS: Those who would ignore, harm, or exploit the disadvantaged. Union-busting companies are a classic example, as are large agricultural firms that exploit farm workers, say, by exposing them to poisonous pesti-cides and paying them poorly.

CATEGORY 3 DEMONS: Those whose activities hurt people or the environment. They include violent criminals and out-of-control police, polluters, those who make unsafe products or engage in consumer fraud, developers with no sense of ecology, and large companies that make extensive profits from government subsidies (e.g., mining, grazing, water, and lumber subsidies) by contributing to the coffers of politicians.

CATEGORY 4 DEMONS: Those who are against public sup-port of education, art, and scholarship.

CATEGORY 5 DEMONS: Those who are against the expan-sion of health care for the general public.

If there is a demon-of-all-demons for liberals, it is Newt Gingrich.

It should come as no surprise that conservative model citi-zens are often liberal demons, and conversely. Now that we know what conservatives and liberals consider basic moral categories, model citizens, and citizens-from-hell, other gen-eral political and social attitudes fall into place.

Incidentally, the theory given here explains many things: why we have the categories of moral actions that we have, why we have the model citizens we have, and why we have the demons we have. The categories for moral actions arise from the metaphors in the moral system. The model citizens and demons arise from the categories of moral actions.

Categories of Policies

The college loan program is illustrative of the great gulf between conservative and liberal moral categories. But it is, in itself, not a very interesting example, since it is not general enough. College loans are not a great issue of our time, the way, say, affirmative action, environmentalism, and abortion are. A more enlightening way to look at the way moral categorization affects public policy is to consider how whole classes of policies fit into the moral categories of liberals and conservatives.

In the next chapter, we will begin to use conservative and liberal moral categories, model citizenry, and demonology to answer the questions we started with about the great issues. Why do stands on the great issues cluster as they do, with opponents of gun control also opposed to social programs, progressive taxation, gay rights, multiculturalism, and abortion, and so on, while proponents of gun control have the opposite views on these issues. What is the logic behind this clustering? And what is the logic that each side uses against the other?

VARIATIONS

As you read through the next several chapters, recall that their purpose is to account for those who (1) have a coherent politics, that is, those who are strictly liberal or strictly conservative, and (2) those liberals and conservatives who share the central model. But many readers either do not have a coherent politics or are not central cases of liberals or conservatives. As a result, many readers will feel, rightfully, that one position or other that I am discussing does not apply to them. The reason, I believe, is that such readers are not prototypical, and I am describing central prototypes. Many readers thus will either fall under one of the variants of the central model, or have some mix of both liberal and conservative political attitudes. The parameters of variation on the central models will be described in Chapter 17, and those variations should account for a great many more readers' views.

The study of variations is a very important part of this project, since the analysis of the central cases predicts that certain ranges of variations should occur. Systematic variations based on fixed parameters of variation are not counterexamples to such a theory; rather they are confirmatory instances.

Part Four
The Hard Issues

10

Social Programs and Taxes

The metaphor of the Nation As Family is part of the conceptual systems of both liberals and conservatives. In that metaphor, the government is a parent. But what kind of parent, according to what model of parenting?

Liberals apply the Nurturant Parent model. Consequently, it is natural for liberals to see the federal government as a strong nurturant parent, responsible for making sure that the basic needs of its citizens are met: food, shelter, education, health care, and opportunities for self-development. A government that lets many of its citizens go hungry, homeless, uneducated, or sick while the majority of its citizens have more, often much more, than these basic needs met is an immoral, irresponsible government. And citizens who are not willing to support such governmental obligations are immoral, irresponsible citizens.

Social programs are also seen by liberals as ways for the government to simultaneously help people (Category 2) and strengthen itself (Category 5). From this perspective, social programs are conceptualized metaphorically as investments—investments in presently unproductive citizens (those who do not pay taxes and who use up government funds) to make them into productive citizens (those who do pay taxes and can contribute to society). The measure of a social program is whether it produces a return on the investment. A social program that doesn't work is a bad investment. The question is not whether to have social programs, but rather which ones work well, that is, which ones produce dividends in the long run.

Liberals also conceptualize social programs as investments in communities. By putting money into the hands of people who don't have it, the government creates jobs in poor communities. People with those jobs spend money, which creates more jobs, and so on. If this is done wisely, there can be a multiplier effect and the result can be a net creation of wealth for the society as a whole. Here the metaphor is one of investing in communities, instead of, or in addition to, investing in individuals. This too is in moral action Category 5.

Liberals also see many social programs as functioning to promote fairness (Category 1). They see certain people and groups of people as "disadvantaged." For historical, social, or health reasons, which are not faults of their own, such people have been prevented from being able to compete fairly in pursuit of their self-interest. Racism, sexism, poverty, the lack of education, and homophobia are seen not only as barriers to empathy and nurturance, but also as barriers to the free pursuit of self-interest and self-development by disadvantaged individuals and groups. For liberals, it is the job of the government to maintain fairness, in the service of both moral self-interest and self-development. Hence it is the job of the government to "level the playing field" for the disadvantaged. This is why liberals support affirmative action.

Conservatives, on the other hand, apply the Strict Father model of parenting to the Nation As Family metaphor. To them, social programs amount to coddling people—spoiling them. Instead of having to learn to fend for themselves, people can depend on the public dole. This makes them morally weak, removing the need for self-discipline and will-power. Such moral weakness is a form of immorality. And so, conservatives see social programs as immoral, affirmative action included.

The myth of America as the Land of Opportunity reinforces this. If anyone, no matter how poor, can discipline himself to climb the ladder of opportunity, then those that don't do so have only themselves to blame. The Ladder-of-Opportunity metaphor is an interesting one. It implies that the ladder is there, that everyone has access to it, and that the only thing involved in becoming successful and being able to take care of oneself is putting out the energy to climb it. If you are not successful, then it is your own fault. You just haven't tried hard enough.

From this perspective, a morally justifiable social program might be something like disaster relief to help self-disciplined and generally self-reliant people get back on their feet after a flood or fire or earthquake. There is a world of difference, from the conservative perspective, between having government help a victim of a natural disaster (who does not have himself to blame for his misfortune) and having government help someone who is merely poor (who, in this land of opportunity, has only himself to blame for his poverty).

In addition, there is a related consideration that militates against social programs in the conservative worldview, what we have called the Morality of Reward and Punishment.

Strict Father morality assumes that it is human nature to be motivated by rewards and deterred by punishments. If people were not rewarded for being moral and punished for being immoral, there would be no morality. If people were not rewarded for being self-disciplined and punished for being slothful, there would be no self-discipline and society would break down. Therefore, any social or political system in which people get things they don't earn, or are rewarded for lack of self-discipline or for immoral behavior, is simply an immoral system. Conservatives see the very existence of social programs as unnatural and immoral in this way.

It is for this reason that any form of socialism or communism is seen by conservatives as immoral, and why, for many conservatives, any social program is seen as a form of socialism or communism. Here is a particularly clear statement of the position, explicitly linking political conservatism with childrearing according to the Strict Father model. The statement is by James Dobson, from the updated version of his classic book, *The New Dare to Discipline* (References, B3, Dobson 1992). Dobson is the country's most influential spokesman for conservative family values among conservative Christians. The quotation comes from a section on the importance of behaviorist principles in raising children.

> Our entire society is established on a system of reinforcement, yet we don't want to apply it where it is needed most: with young children. . . . Rewards make responsible efforts worthwhile. That's the way the adult world works.
>
> The main reason for the overwhelming success of capitalism is that hard work and personal discipline is rewarded in many ways. The great weakness of socialism is the absence of reinforcement; why should a man struggle to achieve if there is nothing special to be gained? This is, I believe, the primary reason why communism failed miserably in the former Soviet Union and Eastern Europe. There was no incentive for creation and "sweat equity." . . .
>
> Communism and Socialism are *destroyers* of motivation, because they penalize creativity and effort. The law of reinforcement is violated by the very na-

ture of those economic systems. Free enterprise works hand in hand with human nature.

Some parents implement a miniature system of socialism at home. Their children's wants and desires are provided by the "State," and are not linked to diligence or discipline in any way. However, they expect little Juan or René to carry responsibility simply because it is noble of them to do so. They want them to learn and sweat for the sheer joy of personal accomplishment. Most of them are not going to buy it (Dobson, *The New Dare to Discipline,* pp. 88–89).

Here Dobson makes explicit the link between Strict Father family values and conservative politics. Social programs subvert human nature. They violate the very thing that, in Strict Father morality, makes morality possible: rewards for discipline and punishment for lack of it. Rush Limbaugh belittled the very idea of national health care as "Rodhamized medicine," after superdemon Hillary Rodham Clinton (References, C1, Limbaugh 1993, p. 171). When he did so, conservatives in his audience understood that he was invoking this view of the immorality of social programs in general.

As we shall see below, the principle of the Morality of Reward and Punishment plays an enormous role in the conservative worldview. The reward side rules out any government distribution of wealth or benefits that is not based on free market competition, and it makes the right to the disposition of private property absolute; the punishment side focuses the criminal justice system on retribution. That is a lot for one principle to do, and as we shall see it is central to a great many conservative stands, aside from social programs.

We can now see clearly why liberal arguments for social programs can make no sense at all to conservatives, whether they are arguments on the basis of compassion, fairness, wise investment, financial responsibility, or outright self-interest.

The issue for conservatives is a moral issue touching the very heart of conservative morality, a morality where a liberal's compassion and fairness are neither compassionate nor fair. Even financial arguments won't carry the day. The issue isn't about money, it's about morality.

President Clinton's Americorps program is a very clear example. It is a double social program: a college loan program and a program to help local communities. The Americorps program allows students to pay off their college loans by working for social programs in local communities.

Since the social programs are immoral for conservatives, so is any program that uses government money to pay for workers in such programs. The government's offer to pay off college loans in this way provides a financial incentive for students to work in such programs. Conservatives see such an incentive as a form of pressure placed by the government on students to engage in an immoral activity. Moreover, paying the students constitutes a second social program, which is doubly immoral.

From a conservative perspective, the students are being coddled through the government's provision of a ready-made way for them to pay off their loans; the disciplined conservative alternative would be for students to have to find jobs for themselves in the workplace to pay off loans. Since the students are not seen as doing honest, productive work in the free market when they work in a social program, they are not seen as earning their loan payoff. And since not every citizen can get loans paid off in this way, getting such a loan at low rates is a form of payment for something unearned. Even worse, from the conservative viewpoint, Americorps gives both students and people in communities the idea that the government and individuals *should* be engaging in such activities—that communities should have people paid by the government to come in and help and that helping in such communities *is* an acceptable form of national service. Americorps, for conservatives, is immoral through and through.

Liberals, of course, have a different moral perspective on social programs. Nurturant Parent morality, applied to politics, makes social programs moral, as we saw above. A double social program—at the same time helping communities and the students who work in them—is doubly moral. And the idea that helping such communities is an excellent form of national service is another plus, which makes it triply moral. That is why it is one of President Clinton's favorite programs.

What we have here are major differences in moral worldview. They are not just differences of opinion about effective public administration. The differences are not about efficiency, or practicality, or economics, and they cannot be settled by rational argument about effective administration. They are ethical opinions about what makes good people and a good nation.

What is at issue in the debate over social programs is the very notion of what morality is and how morality applies to government. There is no morally neutral concept of government. The question is which morality will be politically dominant.

From this perspective, we can see why certain conservative proposals have puzzled liberals. Take, for example, Newt Gingrich's proposal that AFDC children be taken away from their mothers and placed in orphanages. How did this support family values? Or Nancy Reagan's alternative to programs to combat teen pregnancy and AIDS by the distribution of condoms to high school students and clean needles to impoverished drug addicts. The First Lady's proposed solution was not to have such programs, but instead to tell the high school students and drug addicts to "Just say no." Both the Gingrich and Reagan proposals seemed idiotic to liberals, but made sense to conservatives. The reasons should now be relatively obvious.

Orphanages

Why should conservatives have proposed that the children of welfare mothers be put in orphanages, even though orphanages may cost more than giving welfare to help mothers to raise their children themselves. Welfare, as a social program, is immoral under conservative values. How does it serve family values to take children away from the only families they have ever known? If the family values are Strict Father values, the answer is clear. To conservatives the problem is the lack of Strict Father values, beginning with self-discipline. They see welfare mothers as not having those values themselves, and not raising their children to have those values. They see orphanages as institutions that will inculcate those values. They believe that, if the children of welfare mothers are raised to have Strict Father values, then the cycle of dependency, immorality, and lawlessness will stop, and that this will help solve the problems of crime and drugs as well. As to the observation that orphanages impose hardships on children and that the children would be denied their mother's love, the conservative reply is clear: These children need to learn the discipline to overcome hardships and they need to learn Strict Father values more than they need the love of a mother who doesn't teach those values. Orphanages may cost the taxpayer more, but if they contribute to a moral society they are worth paying for.

Just Say No

Nancy Reagan's proposed solution to the problem of drugs was to tell children to "Just say no." That idea made no sense to liberals, who saw drug problems as having to do with despair over social conditions, with peer pressure, and with entrapment into addiction.

But to conservatives whose value system gives priority to Moral Strength, the problem of drugs is the personal lack of the moral strength to just say no. It is a problem of personal values, not of social change or drug treatment centers. The conservative answer to the drug problem is the inculcation of Strict Father values, especially the teaching of self-discipline. People without such discipline, who can't say no, are immoral and deserve punishment. They should be imprisoned for drug use.

This is the same as the conservative answer to teen pregnancy and the spread of AIDS. Don't give out condoms or clean needles, as liberals urge. That just encourages promiscuity. Instead, be tough and teach self-discipline, self-restraint, and abstinence. In a moral system in which morality is correlated with self-discipline and chastity and following societal norms, the moral people won't get pregnant or get AIDS. And the immoral people. . . . Well, they have to learn to be responsible for their actions and they deserve what they get if they don't learn. In the short run some people will get hurt, but in the long run, if a societal standard of behavior is set and adhered to, the nation as a whole will be better off.

IMMIGRATION

Within Strict Father morality, illegal immigrants are seen as lawbreakers ("illegals") who should be punished. People who hire them are just pursuing their self-interest, as they should, and so are doing nothing wrong. From the perspective of the Nation As Family metaphor, illegal immigrants are not citizens, hence they are not children in *our* family. To be expected to provide food, housing, and health care for illegal immigrants is like being expected to feed, house, and care for other children in the neighborhood who are coming into our house without permission. They weren't invited, they have no business being here, and we have no responsibility to take care of them.

From the perspective of Nurturant Parent morality, powerless people with no immoral intent are seen as innocent children needing nurturance. For the most part, illegal immigrants fall into this category.

Illegal immigrants are seen as innocent poor people looking for a better life who are often exploited, for example, when they are lured or brought into the U.S. by employers who are willing to break the law to increase their profit. The stigma of illegality and the enforcement of the law should, in such cases, focus on law-breaking employers.

Illegal immigrants typically perform low-status tasks cheaply that citizens will not do for those wages: farm, sweatshop, and restaurant labor, housecleaning, childcare, gardening, odd jobs, and so on. They are a necessary part of the economy, keeping farm and garment-making profits high and food and clothing costs low. They allow families in the middle class and above to have two-job households by providing housecleaning, childcare, gardening, cheap fast food, and so on. When they do this, they support the lifestyles of better-off people, providing an important service to a great many people. They increase the nation's tax base by permitting middle-class families to have two incomes and allowing many industries to make high profits that are subject to taxation. Out of fairness, they deserve to be compensated for their low pay by having their basic needs guaranteed. Since illegal immigrants historically have become citizens, they should be seen as citizens in the making.

Through the Nation As Family metaphor, they are seen as children who have been lured or brought into the national household and who contribute in a vital way to that national household. You don't throw such children out onto the street. It would be immoral.

Here we can see the Nation As Family metaphor playing a critical and almost direct role in the form of reasoning.

Taxation

Dan Quayle, in his acceptance speech at the 1992 Republican convention, attacked the idea of progressive taxation, in which the rich are taxed at a higher rate than the poor. His argument went like this: "Why," he asked, "should the best people be punished?" The line brought thunderous applause.

It should now be clear why, from the conservative worldview, the rich should be seen as "the best people." They are the model citizens, those who, through self-discipline and hard work, have achieved the American Dream. They have earned what they have and deserve to keep it. Because they are the best people—people whose investments create jobs and wealth for others—they should be rewarded. Taking money away is conceptualized as harm, financial harm; that is the metaphorical basis of seeing taxation as punishment. When the rich are taxed more than others for making a lot more money, they are, according to conservatives, being punished for being model citizens, for doing what, according to the American Dream, they are supposed to do.

Taxation of the rich is, to conservatives, punishment for doing what is right and succeeding at it. It is a violation of the Morality of Reward and Punishment. In the conservative worldview, the rich have earned their money and, according to the Morality of Reward and Punishment, deserve to keep it. Taxation—the forcible taking of their money from them against their will—is seen as unfair and immoral, a kind of theft. That makes the federal government a thief. Hence, a common conservative attitude toward the government: You can't trust it, since, like a thief, it's always trying to find ways to take your money.

Liberals, of course, see taxation through very different lenses. In Nurturant Parent morality, the well-being of all children matters equally. Those children who need less care, the mature and healthy children, simply have a duty to help care for those who need more, say, younger or infirm children. The duty is a matter of moral accounting. They have received nurturance from their parents and owe it to the other children if it is needed. In the Nation As Family metaphor, citizens who have more have a duty to help out those who have much less. Progressive taxation is a form of meeting this duty. Rich conservatives who are trying to get out of paying taxes are seen as selfish and mean-spirited. The nation has helped provide for them and it is their turn to help provide for others. They owe it to the nation. What is punishment and theft to conservatives is civic duty and fairness to liberals.

There are, of course, other ways of conceptualizing taxation, proposals that stand outside of the Strict Father and Nurturant Parent models. These are proposals that come from the business community.

The government is commonly conceptualized as a business. If it is seen as a service industry, taxes can be seen as payment for services provided to the public. Those services can include protection (by the military, the criminal justice system, and regulatory agencies), adjudication of disputes (by the judiciary and other agencies), social insurance (as in Social Security and Medicare and various "safety nets"), and so on.

Under the conceptualization of government as a business that provides services to the public, the questions asked are whether the service is cost-effective and efficient, whether the public is getting the kind of services it wants and needs, and whether the public is willing to pay for the services it wants. If taxes are conceptualized as what you pay for government services, then they are neither punishment nor theft nor civic duty.

One might, at first glance, think that such a conceptualization of government might be compatible with conservative moral views. The reasoning goes like this: Conservatives are pro-business. Why wouldn't they want to see the government operate as a business, in this case a service industry? It would force government to become efficient and cost-effective (see References, D1, Barzelay 1992).

Indeed, President Clinton's "Reinventing Government" program, under the direction of Vice-President Al Gore, has many of these elements. But as Rush Limbaugh would probably say if he got the chance, "A rose by any other name smells just as. . . ." The government may be downsized, streamlined, and made more efficient and cost-effective. It may be de-bureaucratized and made much more responsive to the public. Taxation may be reconceptualized as payment for services. But from the perspective of conservative morality, it is still taxation. It violates the Morality of Reward and Punishment in two ways. First, you don't have a choice as to whether to purchase this service. The government still takes the tax money you've earned, which by the Morality of Reward and Punishment, you deserve to keep. Second, it is still a huge system that does not work by the Morality of Reward and Punishment. It is an enormous system in which the incentive for profit motive does not apply, and the Morality of Reward and Punishment sees such systems as serving the immoral purpose of removing the incentive of reward, the very basis of morality.

You may metaphorically think of the government as a business, and bring principles of good business practice to it, and make it responsive to the public as a good service industry would be, but the government will still not be a profit-making enterprise. That is why conservatives want to privatize government as much as possible. And it is why President Clinton's successes in streamlining government and making it more cost-effective did not earn him high marks with conservatives.

Taxation is not merely *a* moral issue; the very basis of morality is at stake! That is why the issue of taxation is at the very heart of conservative moral politics.

Military Spending

Ronald Reagan came into office pledging to spend less on government. Yet he increased the military budget significantly. Was this a contradiction?

In the summer of 1995, the conservative House of Representatives cut billions out of programs for the poor—$137 million from Project Head Start alone. Yet the conservative House, ostensibly committed to budget cutting, allocated to the military $7 billion more than it had requested. It also supported the reinstitution of expensive and controversial Star Wars research (see References, D2, S. Lakoff and H. F. York, 1989).

Why are conservatives, who say they want to spend less on government, allocating much more to the military than it even requests in inflated estimates? Given that the Cold War is over and we are not in danger of invasion, why do conservatives want to increase military spending, even though it means bigger government?

In the Strict Father model, it is the duty of the strict father to protect his family above all else. By the Nation As Family metaphor, this implies that the major function of the government is, above all else, to protect the nation. That is why conservatives see the funding of the military as moral, while the funding of social programs is seen as immoral.

There is more than a little irony in this. The military is, on the inside, a huge social program, with its own health care, schools, housing, pensions, education benefits, PX discounts, officers' clubs, golf courses, and so on—all paid for at public expense. But the military represents the strength of the nation, and strength has the highest priority in the Strict Father model.

Moreover, the military itself is structured by Strict Father morality. It has a hierarchical authority structure, which is mostly male and sets strict moral bounds. The ethic of moral strength has priority: Everything is keyed to hierarchical authority, self-discipline, building strength, and fighting evils. It is the principal governmental institution that embodies Strict Father morality. Supporting the military as an institution is supporting the culture of Strict Father morality. This makes the military sacrosanct to conservatives. Since it functions in support of conservative morality, conservatives see it as worthy of support even beyond its protective function.

Liberals, focusing on issues of nurturance, see other priorities as more important than the military. They note that the U.S. spends more on its military than the rest of the world combined. Given that we are not in danger of being invaded, and given the end of the Cold War, liberals see no need for much of the military spending. At present, the U.S. is prepared to fight two wars on two fronts, which is seen as overkill. We still maintain 100,000 NATO troops in Europe, which to many liberals is pointless. Much of the money spent on the military could be spent in much better ways, strictly from the point of view of cost-effective government.

But to conservatives, support for the military is support for conservative values. People who go through the military often enter with Strict Father moral values or acquire them. To spend less money on the military is to weaken Strict Father morality—and political conservatism. Correspondingly, for liberals, spending less money on the military means freeing up more for social programs. That, for liberals, is a means to a moral end.

Morality, Not Just Money

Throughout this section of the book, I will be arguing that political policies have everything to do with moral visions— for both liberals and conservatives. The conservative political agenda, for example, is not merely to cut the cost of government. The conservative agenda, as we shall see, is a moral agenda, just as the liberal agenda is.

Consider, for example, the issue of the deficit. How did it get so large?

Liberals like to think of Ronald Reagan as stupid. Whether he was or not, those around him certainly were not. While constantly attacking liberals as big spenders, the Reagan and Bush administrations added three trillion dollars to the national debt by drastically increasing military spending while cutting taxes for the rich. They could count; they saw the deficit increasing. They blamed the increases on liberal spending, but Reagan did not veto every spending bill. Moreover, Reagan's own actions accounted for much of the deficit increase. Had financial responsibility and the lessening of spending been Reagan's top priorities, he would not have allowed such an increase in the deficit, simply by not cutting taxes and not pushing for a military buildup far beyond the Pentagon's requests.

While the deficit was increasing, there was a vast shift of wealth away from the lower and middle classes toward the rich. Liberals, cynically, saw this shift as Reagan and Bush making their friends and their political supporters rich. Certainly that was the effect. It is hardly new for the friends and supporters of politicians in power to get rich. This is usually seen as immorality and corruption, and with good reason. Many liberals saw Reagan that way.

But Ronald Reagan did not consider himself as immoral. Certainly he and his staff could tell that their policies were producing vast increases in the deficit, when they had come into office promising a balanced budget. Reagan was not forced to pursue deficit-increasing policies. Why did he do so?

I would like to suggest that he pursued deficit-increasing policies in the service of what he saw as overriding *moral* goals: (1) Building up the military to protect America from the evil empire of Soviet communism. (2) Lowering taxes for the rich, so that enterprise was rewarded not punished. Interestingly, for President Reagan as for any good conservative, these policies, however different on the surface, were instances of the same underlying principle: the Morality of Reward and Punishment.

What was evil in Soviet communism, for Reagan as for other conservatives, was not just totalitarianism. Certainly Soviet totalitarianism was evil, but the U.S. had supported capitalist totalitarian dictatorships willingly while overthrowing a democratically elected communist government in Chile. The main evil of communism for Reagan, as for most conservatives, was that it stifled free enterprise. Since communism did not allow for free markets (open to Western companies) or for financially rewarding entrepreneurship, it violated the basis of the Strict Father moral system: the Morality of Punishment and Reward.

Adding three trillion dollars to the deficit actually served a moral purpose for Ronald Reagan. It meant that, sooner or later, the deficit would force an elimination of social programs. He knew perfectly well that the military budget would never be seriously cut, and that a major increase in tax revenues to eliminate the deficit would never be agreed upon. In the long run, the staggering deficit would actually serve Strict Father morality—conservative morality—by forcing Congress to cut social programs. From the perspective of Strict

Father morality, Ronald Reagan looks moral and smart, not immoral and dumb as many liberals believe.

The ultimate conservative agenda, as I will be arguing in the following pages, is moral, not financial. It is a thorough political revamping of America in the service of a moral revolution, a revolution that conservatives believe will make Americans better people and improve American life. So far as I can tell, the main issue in every conservative political policy is morality—good versus evil. There is nothing surprising in this. Conservatives consider themselves moral people and they talk about morality and the family constantly. But to liberals, who have their own very different moral system, conservative policies are so immoral that any conservative discussion of morality is taken as demagoguery.

Of course, liberals also see their policies as moral and their overall politics as serving moral goals. Conservatives, however, talk as if liberals were degenerates opposed to morality; as if they were corrupted by special interests; as if they loved expensive and inefficient bureaucracy; as if they wanted to take away the rights of citizens. Each side sees the other as immoral, corrupt, and lunkheaded. Neither side wants to see the other as moral in any way. Neither side wants to recognize that there are two opposed, highly-structured, well-grounded, widely accepted, and utterly contradictory moral systems at the center of American politics.

The failure to see that politics is fundamentally about morality demeans American politics. It makes all politicians look immoral. And it hides the deep logic behind political positions.

11

Crime and the Death Penalty

No topic draws a clearer line between liberals and conservatives than that of violent crime. Strict Father morality sees the cure for violent crime simply as strict punishment. This derives from the Strict Father model of the family that demands that disobedience must be punished, preferably in a painful fashion with an instrument like a belt or a rod. It assumes the Morality of Reward and Punishment, which says that punishment is the moral alternative. And it also assumes a behaviorist theory of human nature that says punishment will work to eliminate violence.

In addition, conservatives claim that violent crime has been the result of "permissive" childrearing practices. They claim that violent crime in later life is caused by a lack of strict discipline at home, a lack of painful corporal punishment in response to disobedience. A mother's nurturance without a father's discipline, they imply, produces antisocial, uninhibited, violent children with no respect for law. Conservatives, using this reasoning, attribute the rise in violent crime to the corresponding decline in the presence of fathers in American homes, due to divorce and illegitimacy. The assumption is that a father would administer strict discipline, with painful corporal punishment for disobedience, and that this would teach children to behave and to grow up as law-abiding, self-reliant citizens.

The Nurturant Parent model of the family makes exactly the opposite claim. It says that children are best socialized and taught responsibility through a nurturant upbringing where discipline is maintained through loving, respectful, and firm interactions and a constant attention to mutual responsibilities and explanations. Painful corporal punishment, the nurturant model says, does just the opposite of what it is intended to do. It teaches violence and violence begets violence. Children who are made to submit through pain to the will of a parent are taught to make others submit to them through their use of violent methods. Correspondingly, neglect has a similar effect. Neglect is a lack of the nurturance in which discipline comes out of loving, responsible interactions. Neglect is thus a form of violence, a denial of needed nurturance.

Liberals respond that violence among fatherless children living in high-crime districts is a result of one or more of the following: (1) mothers who act like abusive strict fathers, administering corporal punishment for disobedience and berating their children; (2) mothers who are neglectful; or (3) social causes, such as poverty or peer pressure. Liberals further argue that mothers who are abusive or neglectful were abused or neglected themselves. The long-term cure for violent crime, liberals argue, is (1) nurturant environments in which there are no neglectful or abusive strict-parent models and (2) the reduction or elimination of poverty by the provision of job training and jobs. From the liberal perspective, what the conservatives are suggesting would just increase violence.

Advocates of nurturant-parent child-rearing practices cite research indicating that strict-father families and corporal punishment contribute importantly to delinquency and violence in later life. Such studies will be discussed in Chapter 21.

Gun Control

Liberal support for gun control is a consequence of the nurturant parent's view of painful corporal punishment—that it contributes to a cycle of violence. Guns are not intended just for target practice or sport. They are intended to hurt or kill people. The very presence of a gun evokes scenarios in which guns are used. These scenarios (self-defense, retribution, or revenge) all share the property that violent punishment is seen as the natural response to wrongdoing. That very idea, the Nurturant Parent model claims, leads to further violence. And further violence with guns means more killing.

Conservatives' support for the right to bear arms—even the right to bear machine guns—comes from Strict Father morality, which says that it is the responsibility of everyone to protect himself as well as he can and it is the responsibility of the Strict Father to protect his family. Guns are seen as the individual's form of protection in a hostile world and they are symbolic of the male role as family protector. They are an instrument of moral strength and a symbol of the power of the Strict Father. As such, they also uphold the moral order. There is thus a very good reason why it is conservatives who support the right to bear arms at a time when conservatives are down in general on rights as liberals have defined them, e.g., the right to a decent standard of living, the right to an education, and the like.

There is also a good reason why very impassioned opposition to gun control often goes with survivalism. Survivalism is about self-reliance through self-discipline, the hallmark of Strict Father morality. And there is a good reason why those who are impassioned about the right to bear arms and about survivalism are also against the income tax. As we have seen, opposition to taxation fits Strict Father morality. And there is a reason why advocates of the right to bear arms are often violently anticommunist. The Strict Father model provides the link between the protective function of the father and the principle of the Morality of Reward and Punishment, the very basis of all morality in the Strict Father model.

This is by no means to say that all conservatives are gun nuts, survivalists, antitax activists, and strong anticommunists. But there is a good reason why those values fit together and why people with those values tend to be conservatives.

Crime

Why do conservatives believe in spending money to build more prisons, and in tougher sentencing laws even for nonviolent offenders. Why do they support the Three-Strikes-and-You're-Out law, mandating twenty-five-year-to-life sentences for repeat nonviolent as well as violent offenders? Why do they do so in the face of evidence that having more people in prison does not reduce crime?

The state of Minnesota's Kids First program, which stresses day care, education, and community involvement, has succeeded in crime prevention at a much lower cost than running prisons. Why has this model not been supported by conservatives?

The answer comes out of Strict Father morality, in which the Moral Strength system of metaphors is primary, with Moral Self-Interest right behind it. Strict Father morality thus includes Retribution, Moral Strength, Moral Self-Interest, and Moral Essence. Retribution sees punishment as defining justice. The priority of Moral Strength entails that a show of strength is the best protection against evil. It is a consequence of Moral Self-Interest that people act in their own self-interest; hence, people will commit crimes if it is in their interest (that is, if punishment is lenient) and won't commit crimes if it is not in their interest (if punishment is harsh). And according to Moral Essence, past behavior is a guide to essential character and essential character predicts future behavior. Therefore, a repeat offender has a bad character, which means he's likely to commit crimes again. To protect the public, he should be imprisoned for a long time.

By Strict Father morality, harsh prison terms for criminals and life imprisonment for repeat offenders are the only moral options. Programs like Minnesota's Kids First are social programs and are, as such, immoral to conservatives for reasons given above. The conservative arguments are moral arguments, not practical arguments. Statistics about which policies do or do not actually reduce crime rates do not count in a morality-based discourse.

Liberals, following Nurturant Parent morality, point to Minnesota's Kids First as an argument that prevention programs can reduce crime, while pointing to statistics indicating that putting people in prison does not. Liberals see crime as having social causes—poverty, unemployment, alienation, and lack of caring and community—and argue that social programs are needed to address those social causes. Conservatives don't believe in social causes of crime or in any other social causes. Let's consider why.

Class and Social Causes

Conservatives tend not to use explanations based on the concepts of class and social causes, nor do they recommend policy based on those notions. Why? Liberals use these concepts all the time, in providing explanations and in formulating and justifying policy. Again, why? What is it about the difference between liberals and conservatives that makes these concepts sensible to one, but not to the other?

Think for a moment about how the notions of class and social forces are used. Class structure comes with the notion of an upper class with wealth and power that wants to maintain its privileges; a lower class—the cheap labor of the upper class—held down and kept subservient to the upper class; and a middle class caught in between—aspiring to the upper class and afraid of falling into the lower class, but also depending on the cheap labor of the lower class. Social forces are usually postulated to account for the failure of certain lower-class groups to succeed, to gain access to wealth and power. People in the lower class can get caught in the system and be unable to rise. Such a social arrangement is seen by liberals as unfair. It is a social injustice.

According to this picture, the upper and middle classes could not maintain their current lifestyles without the cheap—and often difficult and demeaning—labor of the lower class: picking vegetables, working in fast-food places, cleaning houses, collecting garbage. In this picture, the upper classes owe a lot to the lower class—much more than they are paying. Social justice demands that the lower class be paid more, live under better conditions, and be given maximal opportunities to work their way out of poverty, opportunities for education and job training, for example.

This picture is usually supplied as a justification for government to do something to help out the people at the bottom, at least to provide for their basic needs and to give them enough education and job training to allow them to do a bit better. It is also used to explain the rage and violence of lower-class people against the system that "imprisons" them socially and economically. It is the class structure and the social forces holding it in place that does the "imprisoning."

Concepts like "class" and "social and economic forces" and "social and economic imprisonment" fit naturally into a liberal worldview. For liberals, the essence of America is nurturance, part of which is helping those who need help. People who are "trapped" by social and economic forces need help to "escape." The metaphorical Nurturant Parent—the government—has a duty to help change the social and economic system that traps people. By this logic, the problem is in the society, not in the people innocently "trapped." If social and economic forces are responsible, then other social and economic forces must be brought to bear to break the "trap."

This whole picture is simply inconsistent with Strict Father morality and the conservative worldview it defines. In that worldview, the class hierarchy is simply a ladder, there to be climbed by anybody with the talent and self-discipline to climb it. Whether or not you climb the ladder of wealth and privilege is only a matter of whether you have the moral strength, character, and inherent talent to do so. Because explanations for success or failure give priority to Moral Strength and Moral Essence, explanations in terms of social forces and class make no sense. They are only seen as excuses for lack of talent, laziness, or some other form of moral weakness. In such a worldview, the concept of social justice does not make sense. If the poor are selling their labor to the rich, then it is the labor market and the labor market alone that determines what that labor is worth. Labor, in this metaphor, is a commodity like any other commodity, and its value is not inherent but determined by what people are willing to pay in exchange for it. The Morality of Reward and Punishment, which requires that all markets be free markets, demands this. Any other arrangement would be immoral and threaten the very moral foundations of society. It is for this reason that conservatives are against the minimum wage, while liberals, in the name of a bare minimum of social justice, support it.

To conservatives, the existence of a wealthy class simply

makes real the Morality of Reward and Punishment, the basis of all morality. It is not wrong, not something to be corrected, for the wealthy to seek further privilege. It is natural and moral, a guarantee that the Morality of Reward and Punishment continues to work. Crucial to this is what conservatives see as the essence of America—the Ladder of Success myth. As long as free enterprise flourishes and anyone with enough self-discipline and imagination can become an entrepreneur, the Morality of Reward and Punishment will hold and all will be well.

The logic of conservatism locates so-called "social" problems within people, not within society. For this reason, it would make no sense to conservatives to use class and social forces as forms of explanation and justification for social policy.

Nature and Nurture

The term "nurture" has two related senses, one having to do with nurturance (nurture-1) and the other with environmentally determined rather than genetically determined factors in human development (nurture-2), as in the opposition between "nature" and "nurture." Nurturance is certainly an environmentally determined factor, and it leads liberals to look almost exclusively at environmentally determined factors in social and political explanations.

Though Strict Father morality is opposed to the goal of a nurturant society, it is extremely concerned with environmental factors, such as childrearing and the more general use of reward and punishment. But there are other parts of Strict Father morality that are concerned with nature as opposed to nurture-2.

Strict Father morality gives high priority to Moral Essence and the idea that the Moral Order is the *natural* order of dominance. Thus, there is a strain in conservatism that uses nature, as well as environmentally determined factors, as a means of explanation for social problems. Conservatives can have it both ways. Unsuccessful people can fail for one of two reasons: because they lack either (1) character (which is environmentally determined) or (2) talent (which is natural). That is why conservatives tend to like books like *The Bell Curve*, while liberals tend not to. *The Bell Curve* provides the second explanation for the economic failure of blacks— lack of talent; but if that can't be proved, there is always lack of character as an explanation.

Preventing Crime

The difference in conservative and liberal moral systems leads to different views of the role of nature and nurture and of the explanatory validity of concepts like class and social forces. To conservatives, any appeal to social forces is always just an excuse for lack of talent (nature) or lack of character (nurture-2). Conservatives therefore don't address crime by looking for social causes. They address crime just the way they would address the refusal of a child to abide by his parents' rules—according to the Morality of Reward and Punishment. Conservatives punish crime, and they assume that if crime is punished harshly enough, it will end, because criminals will have a strong enough disincentive to keep them from committing crimes. If disincentives don't work, then the criminals must be inherently bad—rotten to the core—and should be locked away for life or for a very long time.

Given the central position that the Morality of Reward and Punishment has in the conservative moral system, it is no surprise that conservatives in most cases prefer retribution over restitution as a form of justice, as a way of balancing the moral books. It is therefore no surprise that conservatives are in favor of the death penalty. It is a form of retribution, a life for a life.

Liberals, of course, look at these issues very differently. The primacy of empathy leads to an overriding concern with fairness toward anyone committing a crime (Moral Action Category 1). Given the overwhelming power of the state, any citizen is helpless by comparison unless the rule of law is carried out scrupulously. Great care must be exercised that people accused of crimes be given a fair trial and that their rights as citizens not be overridden by the state. Liberals are acutely aware that the police can abuse their power and act unfairly to get a conviction. This is especially true where the accused is someone who is poor or a member of a minority group. Poor people cannot get as good representation in court as rich people, and members of minority groups are subject to prejudice. Most of the people who have been given the death penalty and are on death row are poor members of minority groups. Overwhelming concerns for empathy and fairness, which grow out of the Nurturant Parent model, lead liberals to have a paramount concern for the abuse of state power.

Conservatives see such a concern as coddling criminals, as caring more about the criminal than the victim. They are puzzled by liberal behavior. If liberals are so concerned about protecting the helpless, why aren't they more concerned about protecting the victims of crime? Why aren't they promoting stricter penalties in the name of protection?

It is not that liberals are not concerned about the victims of crimes. Rather, they disagree about how crime is to be minimized overall. First, the rule of law must be upheld. If the state can act like a criminal, framing innocent people and trampling on the rights of the accused, then all hope for the rule of law is lost. To keep the state and its representatives— the police and the courts—honest, the rights of everyone accused of a crime must be upheld strictly.

Fairness is the issue here. If the legal system is not fair, it will have no legitimacy. Fairness is a consideration that arises, in Nurturant Parent morality, out of concerns for empathy and nurturance for all.

Second, liberals do believe in social causes; they believe that if children are not raised in nurturant environments, they will not learn to behave responsibly toward others. If a gang is the closest thing to a nurturant community a child knows, then he will engage in gang behavior. Thus, the way to best inculcate responsible behavior over the whole society is to provide nurturant environments for as many people as possible. The best long-range approach to fighting crime is to fund programs like Project Head Start, to provide high-quality day care for poor working parents, to provide high-quality education for the poor, and so on. That is why a liberal like Anthony Lewis (*New York Times*, August 8, 1995) is so outraged when he sees conservatives cutting $137 million from Project Head Start and simultaneously giving the military $7 billion more than it said it needed. It costs more than $20,000 a year to house one prisoner for a year—the cost of tuition at an Ivy League college. Liberals argue that, in the long run, it is more effective, a lot cheaper, and a lot better for everyone to fund Project Head Start, day-care centers, and so on. But conservatives believe that there can be no such thing as social causes of crime, that crime is always a matter of individual moral weakness. It follows then that it is silly to spend money on countering nonexistent social causes.

Third, liberals do not believe in the overriding Morality of Reward and Punishment. They do not believe that it is mainly fear of punishment that binds society together and makes people act kindly, responsibly, or at least civilly, toward each other. Liberals believe that it is nurturance that brings this about, that loving and supportive parent-child bonding creates communities in which there are strong social ties. It is not that rewards and punishments are never appropriate, but they are not the basis of morality, nurturance is.

Simply increasing penalties, liberals argue, will not eliminate crime. The death penalty, liberals argue, has been shown not to be a deterrent to murder. Murderers apparently do not do a cost-benefit analysis before killing someone, and so the death penalty does not deter them.

The Death Penalty

The death penalty itself is a major dividing line between liberals and conservatives. Supporters tend to be conservative and opponents tend to be liberal. Nurturant Parent morality militates against the death penalty. The major legal argument given rests on fairness, which, as we have seen, arises in Nurturant Parent morality from considerations of empathy and nurturance. The argument is that the courts do not have a way to guarantee that the death penalty is applied fairly. Prejudice and politics do enter into murder trials. If a person unfairly convicted of murder is executed, then there is no recourse if he is later discovered to have been innocent. Most of the people on death row are poor and black and are unable to afford adequate legal representation, which, it is argued, makes it more likely that they will get the death penalty. The same penalties should apply to all racial and economic groups. If that is not true of the death penalty, the death penalty should not be applied at all.

But liberals' feelings about the death penalty run much deeper than that. Nurturance itself implies a reverence for life, which is a form of the unconditional love of nurturant parents for their children. If the government is conceptualized as a nurturant parent, then it should have such an overwhelming reverence for life itself. The death penalty denies such a reverence for life, and so is inconsistent with the conception of the government as nurturant parent.

In Strict Father morality, the strict father metes out punishment for the wrongdoings of children. The Nation As

Family metaphor makes the government into the Big Daddy who is meting out punishment. Are there any limits on the harshness of punishment? The lack of such limits in the family would make it moral for a parent to kill his child in the name of discipline. When parents are abusive, this can happen, and all too frequently does. Infanticide is what the death penalty amounts to in the Nation As Family metaphor. The spectre of the state functioning like a murderously abusive parent is, I believe, what lies behind the constitutional prohibition against cruel and inhuman punishment and behind liberals' abhorrence of the death penalty.

For these reasons, I believe, liberals find the death penalty uncivilized and point out that nations around the world have banned it. It sets a bad example for the state to be engaged in killing people, the worst crime someone could commit. Just as parents can set bad examples, so, according to the Nation As Family metaphor, can the government. Governments should not be in the business of killing people.

Arguments against the death penalty are not just about the death penalty itself. They are emblematic of a broader issue. They are about how the state should be conceptualized and how it should function in general.

The liberal arguments, of course, cannot possibly be persuasive to conservatives. If the very basis of morality is reward and punishment, then in a moral society the way to deal with crime is punishment, an eye for an eye—period. The argument that the death penalty does not deter murder doesn't really matter that much. To conservatives, the death penalty is part of a moral society, in which the Morality of Reward and Punishment rules supreme.

12

Regulation and the Environment

A major aspect of nurturance is protection. Parents have to protect their children not just from the most obvious forms of crime and violence but from less obvious forms of danger: cigarette smoke, asbestos and other toxic chemicals, lead paints, dangerous toys, harmful foods, cars without seat belts, inflammable clothing, dangerous toys, unscrupulous businessmen, and on and on. A truly caring, nurturant parent is vigilant when it comes to such everyday dangers.

Government, seen as a nurturant parent, must similarly be vigilant about such everyday dangers to its citizens. Liberals conceptualize government regulation as the protection of those who cannot protect themselves—protection of citizens, workers, honest businessmen, and the environment against possible harm by unscrupulous or negligent businesses and individuals. Government regulation of business is there to be sure that businesses don't hurt or cheat anyone. Long experience tells us that citizens need such protection. Unscrupulous or careless businesses in America have a long history of putting their workers in danger, polluting the environment, producing dangerous products, and cheating their customers. The job of government regulation is to minimize this.

Conservatives, however, do not conceptualize government regulation as protection. Why? Given conservative moral priorities and moral categories, they could not possibly conceptualize government regulation as protection. Who could it be protection against? Certainly not model conservative citizens, successful businessmen, from whom we have nothing to fear. In the system of conservative moral categories, government regulation falls under interference with the pursuit of self-interest by people trying to make a living, people using their self-discipline to become self-reliant (and, if possible, rich). These are the good people in our society. We want to encourage people like them and it is wrong to put roadblocks in their way.

The argument against environmental, worker-safety, and product-safety regulations is that they are too cumbersome and get in the way of doing business. Regulators are seen as stupid and corrupt. But conservatives don't just want to reform regulatory agencies, to get rid of what is cumbersome, while still protecting those who need protection. They want to get rid of regulations completely. Liberals are incredulous. Don't people have a right to clean air and water, safe products, safe airlines, safe jobs? Shouldn't the environment be preserved for our grandchildren and great-grandchildren? The liberal arguments never get heard by conservatives. They can't. The primary conservative moral categories filter out the liberal arguments. The primary conservative moral categorization, which is the political version of Strict Father morality, makes conservative businessmen into model citizens and regulation into interference with them.

The basis of the classification of successful businessmen as model citizens is very deep, as we have seen. It is the principle of the Morality of Reward and Punishment, which is at the very heart of Strict Father morality. To place restrictions on that principle is to strike at the very heart of conservative ethics and the conservative way of life. Placing restrictions on moral people who are engaged in moral activities is immoral. That's why conservatives see government regulation as immoral. Once successful businessmen are categorized as model citizens, there is no possibility of seeing regulation as protection.

The Environment

Strict Father morality includes the notion of the natural order of domination: God has dominion over human beings; human beings over nature; parents over children; and so on. By the metaphor that the Moral Order Is the Dominant Order in nature, this arrangement is seen as a moral one, with the responsibility for protection and care going along with moral authority. But man's protection and care make sense only in the context of the primary given: the domination of man over nature according to man's priorities, which is taken to be both natural and moral. Newt Gingrich says all this straightforwardly:

> For me, any such effort [at environmentalism] begins with the premise that man dominates the planet and that we have an absolute obligation to minimize damage to the natural world. I am not a preservationist. It is impossible for us to be a dynamic species and still act as if we don't exist. (References, C1, Gingrich, p. 195)

It is assumed that being a "dynamic species"—one that figures out how to get what it wants and then goes for it—is naturally going to result in "damage" to the environment. You don't stop going after what you want, but you do try to minimize the damage.

Strict Father morality starts with the "natural order," the

domination of man over nature. It adds the Morality of Self-Interest, in which the self-interest of all is maximized if we each seek our personal self-interest. It further adds the Morality of Reward and Punishment as the basis of morality, which implies that it is immoral to stop individuals from working hard for the sake of profit.

What all this adds up to is the view that nature is there as a resource to be used by man for his self-interest and profit. But, frugality being a virtue, the resource should be "conserved" as much as possible. We should use it for our individual aims but not be too wasteful. Conservative environmentalism is "conservation"—"hardheaded management with regard to costs and benefits" (Gingrich, *To Renew America*, p. 198). "To get the best ecosystem for our buck, we should use decentralized and entrepreneurial strategies, rather than command-and-control bureaucratic effort" (ibid., p. 196). In short, there should be no attempts at top-down overall controls on what happens to an ecosystem. Our relationship to nature should work according to free-market principles.

Conservative environmentalism—the conservative view of man's relationship to nature—arises naturally from Strict Father morality. It is conceptualized and reasoned about in terms of a system of common metaphors that best fit the Strict Father view. Those metaphors are:

- Nature Is God's Dominion (given to man to steward wisely).
- Nature Is a Resource (for immediate human use).
- Nature Is Property (for the use of the owner, and for sale and purchase).
- Nature Is a Work of Art (for human appreciation).
- Nature Is an Adversary (to be conquered and made to serve us).
- Nature Is a Wild Animal (to be tamed for our use).
- Nature Is a Mechanical System (to be figured out and put to use).

These modes of metaphorical thought allow us to comprehend and reason about nature in accord with what Strict Father morality tells us.

The stewardship metaphor says that nature, by God's authority, is man's to use for whatever he wants, but he should use it sensibly and frugally.

The resource metaphor (consider the term "natural resources") assumes that whatever is in nature is, and should be, part of a human economic system. Its value is not intrinsic but is determined by how useful it is to human beings and how plentiful it is. If it is plentiful, then by the law of supply and demand, its value will be low. If it is rare, its value will be high. It also follows that certain aspects of nature will be categorized together in classes determined by human purposes. Salmon, for example, is classified in our economic system as a fish used for food. Its value is its use as a food. If salmon become extinct, it is no big deal because there are still many other fish that can be used as food, that serve the same function as a resource. The mere existence of salmon living in streams and spawning there is in itself of no value in the metaphor of nature as a resource.

The property metaphor is, of course, not universal. There are cultures where the very idea of a person being able to own a tree or a forest or a mountain would be ludicrous. Under the property metaphor, nature is up for sale, just like a sofa or a car or a video game. Things in nature—forests, lakes, volcanoes, canyons—can be put to any use the owner wants, or even destroyed if that is the owner's wish. The value of nature, in this metaphor, is as a commodity, and it fluctuates with local tastes and market conditions. In this metaphor, streams, lakes, and valleys have no inherent value, only market value.

The work-of-art metaphor assigns nature an aesthetic value dependent on human aesthetic sensibilities. A Grand Canyon or a Yosemite or a whale takes on value because of how it is aesthetically judged. A desert ecology or a homely newt will, for most people, have low aesthetic value.

The adversary metaphor makes the "conquest of nature" noble and man's domination of nature something worked for and therefore earned and deserved. It assumes an alienation of man from nature, a separateness that can be overcome only through domination. It assumes that nature, in an unconquered state, is dangerous to man and that man must dominate nature to survive. The continuing and expanding domination of nature becomes, by this metaphor, a form of self-defense and a moral enterprise.

The wild-animal metaphor again sees nature as alien to man and dangerous, but of possible economic use. The "taming" and domination of nature thus becomes a noble enterprise—and a profitable one.

The mechanical-system metaphor characterizes our understanding of the role of science vis-à-vis nature. Science is seen as having the job of figuring out what makes nature tick and what nature's internal workings are. The purpose is control, so that we can use nature for our purposes.

Nurturant Parent morality sees our relation to nature in a very different light. The natural world is what gives us life, what makes all of life possible, and what sustains us. Nature has provided and continues to provide. Our relationship with nature is as the recipient of nurturance and as such it involves attachment, inherent value, gratitude, responsibility, respect, interdependence, love, adoration, and continuing commitment.

This view of nature is conceptualized using the following metaphors.

- Nature Is a Mother (who provides for us).
- Nature Is a Whole (of which we are inseparable parts).

- Nature Is a Divine Being (to be revered and re-spected).
- Nature Is a Living Organism (whose needs must be met if it is to survive).
- Nature Is a Home (to be maintained and kept clean).
- Nature Is a Victim of Injury (who has been harmed and needs to be healed).

The nature-as-mother metaphor sees nature as nurturer and provider. The normal relationship of a child to a nurturant mother is one of attachment and love, something of continuing inherent value that cannot be bought and sold—a relationship that gives meaning to one's life. One's moral attitude to a nurturant mother is one of gratitude, responsibility, and respect. You have a responsibility to provide for her needs to the best of your ability. You accord her dignity. You show your gratitude in your deeds. The relationship is one of mutuality, of interdependence. And it is not a weak or temporary relationship, it is a continuing commitment.

The whole-part metaphor stresses the attachment, mutuality, and interdependence of the relationship.

The divine-being (Earth As Goddess) metaphor focuses on our dependence, our respect, and our adoration.

The living-organism metaphor (as in the Gaia hypothesis) focuses on interdependence and the fact that ecosystems have needs that must be met if they are to survive.

The home metaphor stresses that the earth is where we live, that it is finite, that it is a place of nurturance and security, that it has to be maintained, that it has to be kept clean. It also focuses on the fact that we are attached to our homes, that there is something about our home that has inherent value, that a home has value far beyond its market value.

The injury metaphor focuses on the fragility of nature and the harm it has already undergone. It implies that nature cannot sustain continuing damage if the natural world as we know it is to survive, and that healing is necessary.

These are the metaphors for nature given highest priority in Nurturant Parent morality. Some of the metaphors used in the conservative view of nature also fit Nurturant Parent morality, but they have subservient status in the system and their meaning is radically changed by that status.

Take, for example, the resource metaphor. This moral system assumes not the domination of man over nature, but the interdependence of man and nature. Nature is a resource to someone nurtured by nature, but the nurturer must be nurtured too—we must take care of the earth. This entails the notion of sustainability. Nurturance requires interdependence if it is to be sustained; the same is true of our relationship to nature. The resource metaphor, in the context of the nurturance metaphor, implies sustainability, a central ecological concept. Yes, nature is a resource, but it must be a sustainable resource. Yes, natural resources have economic value, but not just economic value. Their economic value must be determined within the context of the system of inherent value I have just described.

Since nurturance is an aesthetic experience, the work-of-art metaphor applies, but in a very different way. In nurturance, the aesthetic experience is not separate from the experience of nurturance. Thus, the aesthetic values are not separate from all the other values entailed by nurturance, e.g., respect, attachment, continuing commitment. The aesthetic value of nature is, therefore, not just seeing a pretty scene when you take a walk in the woods. It is not like having a picture of the woods on the wall. The aesthetic experience is part of nurturing.

The mechanical-system metaphor, too, takes on a different meaning in Nurturant Parent morality. Discovering important things about a nurturant relationship that you are in is not separable from the nurturance itself. In a moral system where nurturance is the highest value, discovery is in the service of nurturance. If you live by Nurturant Parent morality, then scientific discoveries are nurturant values in themselves. They also take on importance when they are in the service of nurturance, for example, when they cure us of illness or allow us to communicate with friends and relatives who live far away.

From the perspective of nurturance, the stewardship metaphor takes on a very different meaning. Stewardship is not in the service of domination, but in the service of healing and sustaining nature so that it can continue to nurture us and so that we can continue to reap all the benefits of being in a two-way nurturant relationship with the natural world.

Given these very different moral conceptions of the relationship between human beings and nature, it should come as no surprise that liberals and conservatives have many opposing views on environmental policy. Take the Environmental Protection Agency. From a liberal perspective, the mission of the EPA is to serve the protective function of nurturance. Its function is both to protect the citizenry from environmental dangers and to protect the environment itself. It does its job through the enforcement of environmental laws and through the monitoring of environmental indexes, for example, the monitoring of air pollution. Part of its job is to regulate industries that produce pollution, like steel mills and power plants, to make sure they conform to environmental standards. Another part of its job is to enforce the Endangered Species Act. Still another part of its job is to study how to protect ecosystems, for example, wetlands. Once laws are passed, it decides how best to implement them.

From a liberal point of view, such an agency is absolutely needed. A part of "healing nature" is cleaning up our rivers and streams so that fish can live in them and plants can thrive on their banks, and so that they can be used as much as possible as sources for drinking water. This serves the purpose of both nurturing nature as well as nurturing people by supplying clean water to drink. Since the EPA was formed, its regulations have played a major role in cleaning up sources of drinking water. Liberals are aghast that conservatives have proposed legislation to cut the EPA's enforcement power so drastically that it could no longer stop industries from polluting drinking-water sources.

Conservatives want to eliminate the EPA on the grounds that government should encourage and reward free enterprise, not constrain and punish it. This fits the conservative view of nature as being there to serve private interests. Solutions to as many problems as possible should come out of free enterprise solutions that reward self-discipline and enterprise, rather than out of government-imposed constraints that punish enterprising companies through regulation.

Or take the issue of old-growth forests and the spotted owl. Old-growth forests are not merely many hundreds of years old. They are remarkable ecosystems with a unique variety and large number of smaller plants, insects, birds, and animals that fit together in an extraordinary complex ecosystem like no other. Ninety percent of America's old-growth forests have already been destroyed by logging; only ten percent are left and logging companies want to cut them down. They can never be replaced.

Nurturant Parent morality imposes a view of nature that assigns inherent value to old-growth forests. Their value is not in what you can sell their lumber for, or in how pretty they look, but in what they are. They are a unique form of nature that has almost been destroyed; a nurturant morality demands that they be protected.

Now it happens that there is no law on the books to protect old-growth forests for their own sake. But there is a law called the Endangered Species Act, which forbids the destruction of the habitat of a species that is on the EPA's list of endangered species. As it happens, there is an endangered species on that list that lives in old-growth forests and only old-growth forests: the spotted owl. Save the spotted owl and you save old-growth forests. Of course, the Nurturant Parent view of nature also assigns inherent value to endangered species, and hence to the spotted owl. But there is far more involved here than just the spotted owl. Liberals, naturally, are for the strict enforcement of the Endangered Species Act, and the consequence of saving the remaining old-growth forests.

Conservatives are furious and they should be, given their worldview. The most central features of Strict Father morality are being challenged. The Morality of Reward and Punishment, the very basis of Strict Father morality, is being challenged in two forms. The first is a challenge to unrestricted free enterprise, the unrestricted freedom to seek private reward for your labor and investments. If investors in the logging companies cannot realize expected profits on their investments, then that freedom has been limited. The second is a challenge to the unrestricted use of private property, the unrestricted freedom to do whatever you want with the fruits of your labor, namely, your property. Logging companies own some of these old-growth forests. If they cannot cut them down for profit, then restrictions have been placed on the use of private property. In addition, the very Moral Order has been challenged: the idea that man, by God's will, has dominion over nature and that nature is a resource for man's use. Loggers—ordinary, hardworking, moral, law-abiding folks—may lose their source of income. What is more important, people or owls? Common sense— the common sense of Strict Father morality—says people!

One of the common arguments the conservatives use against environmental regulation is that it can be done better by the use of the market. One proposal is the sale of pollution rights. Ailot to each industry the right to pollute a certain amount. Then, if it pollutes less than that amount, allow it to sell its unused portion to another industry. That gives industries an incentive to pollute less so that they can sell their pollution rights. Done cleverly, its advocates claim, this market system can reduce industrial pollution.

Perhaps so. But that would not address the liberals' environmental concerns. Liberals conceptualize nature in very different metaphors than do conservatives. The liberal metaphors assign intrinsic value to aspects of nature; the market does not and cannot. Market solutions may help in many cases but they cannot address the deepest of liberals' concerns.

When one steps back for a moment and takes a look at conservative and liberal moral systems, it becomes apparent that the issue is not people versus owls or market forces versus the EPA, but two utterly opposed moral visions of the proper relation of man to nature. The issue is a huge one: Which moral view is to dominate, not just on environmental questions, but overall in our culture and our politics?

13

The Culture Wars:
From Affirmative Action to the Arts

Strict Father morality and Nurturant Parent morality are opposed moral systems; they define incompatible moral worlds. Conservatives have understood very well that their goals are not just political and economic. Conservatives want to change American culture itself. They want to change the idea of what counts as a good person and what the world should be like. Conservatives understand that this means starting with the family. But it also means changing such things as who gets what jobs and what ideas dominate our culture. Here is how the Strict Father vs. Nurturant Parent dichotomy plays out over a range of cultural issues, from affirmative action to the nature of art.

Affirmative Action

Strict Father morality comes with a notion of the right kind of person—a self-disciplined person, one who can set his own plans, make his own commitments and carry them out effectively. It requires that competition between people not be impeded in any way if they are to continue to have the incentive to be self-disciplined. Any policy that gives people things they haven't earned is seen as immoral, because it lessens the incentive to be self-disciplined. From this perspective, affirmative action looks immoral to conservatives, on the grounds that it gives preferential treatment to women and minorities. It is a relatively direct consequence of the Strict Father model.

The Nurturant Parent model gives the opposite answer. It is the job of a nurturant parent to see that the children in the family treat each other fairly. In the Nation As Family metaphor, that becomes: It is the job of the government to see that its citizens treat each other fairly. Thus it is the responsibility of the government to guarantee fair treatment of people who have been subject to discrimination—women, nonwhites, and ethnic minorities.

In a nurturant family, the issue of fair distribution concerns the whole family over its whole existence. Where unfairness has existed in the past, some unfairness in the present may be needed to balance things out and make things fair overall. The Nation As Family metaphor makes that true of a nation.

Liberals further adopt the common metaphor that a natural group is an individual, the metaphor that defines collective action and collective rights. It allows considerations of fairness to individuals to apply to groups. Thus, one must look not just at fairness to individual women at the present, but at fairness to the group of women considered as a unit, taking into account both the past and the present.

The use of the Group As Individual metaphor is not arbitrary. Liberals offer two reasons for such a group focus. First, there is the phenomenon of stereotyping. People commonly reason in terms of stereotypes, judging all the members of a class in terms of a stereotypical image. That image is usually based on a past cultural model. For example, women may still be seen fundamentally as housewives whose skills are best suited to homemaking, or who are unable to do rigorous logical thinking, or who lack physical stamina. Present individual women are likely to be, unconsciously, judged that way because of the persistence of stereotypes in the unconscious conceptual systems of both men and women in our culture. This could result in a woman being judged less qualified than a man without the person doing the judging even being aware of his prejudice. The Group As Individual metaphor helps remedy prejudices, whether conscious or unconscious, by measuring fairness with reference to a group over time. Affirmative action is a means for remedying unfairness for whole groups over time.

Does this necessarily involve being unfair to individuals at present? It may, but it may very well not. First, existing unconscious stereotypes create an unfair situation at present. A stereotype helps white males start with an advantage. Another stereotype puts equally or better qualified women or nonwhites at a disadvantage. Affirmative action can help make things closer to even, giving a chance to a better, or equally well qualified, woman or nonwhite, when they would otherwise be eliminated because of the white males' stereotype advantage.

In addition, there is another liberal motivation of the Group As Individual metaphor. Different groups have different subcultures, with different value systems. White men have a different subculture with different forms of discourse and a different value system than white women. The same goes for other cultural groups. At present white men hold top positions throughout our society. They are the ones doing the judging. As honestly as they may do that judging, the values by which they judge will almost inevitably be the largely unconscious values of their own subculture. For example, men may not value certain skills that women have and men tend to lack. As a result, their ''honest judgment'' may be discriminatory. Treating women as a group and measuring fairness by the group over time is one way of overcoming the implicit advantage that white men have—the subculture advantage.

Under affirmative action, white men still have advantages they don't even know they have: stereotype-advantages and subculture-advantages. Affirmative action has not overcome these advantages, and it will take affirmative action over a much longer time to overcome them.

Gay Rights

Why are liberals for gay rights? The answer is simple and straightforward. For liberals, gay rights follow naturally from Nurturant Parent morality. A nurturant parent treats his or her children fairly and loves them equally. By the Nation As Family metaphor, the government, as metaphorical parent, should treat all citizens fairly and equally, gay or not.

Why are conservatives against gay rights? Why is there so much hostility against homosexuals on the part of conservatives? This has nothing to do with disliking big government and bureaucracy, or supporting fiscal responsibility, or supporting states' rights. The answer is Strict Father morality. Gay and lesbian couples simply do not fit the Strict Father model of the family. Homosexuality challenges the monolithic authority of the father. And above all, it challenges the natural order, which presupposes that sex is heterosexual sex in which men are dominant over women and that, in a family, this natural order carries over to the moral order.

But this is not just a matter of the family. The family, conservatives understand well, is the basis of all morality, all social arrangements, and all politics. Homosexuality challenges the very idea that the Strict Father family is the right model of the family, and therefore of morality and politics.

That is why conservatives resist seeing homosexuality as

natural for a certain percentage of the population. Conservatives do not talk much about the increasing evidence that
homosexuality has a genetic basis. Gays speak of "discovering" that they are gay, rather than "choosing" to be
gay. Conservatives, however, speak of the gay "lifestyle,"
as though being gay were simply a conscious choice of a
particular way of life. If there is no choice about being gay,
if one is born gay or bisexual or heterosexual, then the force
is taken away from the idea of homosexuality as an immoral
choice of "lifestyle." Indeed, if free will is taken away, if
there is no choice, then it is much harder to make homosexuality a moral issue.

The conservative version of the Moral Strength metaphor
requires that sexual morality be a matter that one has control
over: it is a matter of self-discipline. If homosexuality is
genetically determined and, therefore, natural, normal, and
out of the domain of free will, the concept of Moral Strength,
which requires that all immoral behavior be preventable
through self-discipline, becomes inapplicable. You can no
longer say: if you just try hard enough, you can be heterosexual. Because the priority of Moral Strength is so central to
the conservative moral system, conservatives will necessarily
have a very hard time accepting the idea that homosexuality
is biologically determined, natural, and normal for a certain
segment of the population.

Interestingly enough, many conservatives would still find
homosexual sex, gay households, and gay families immoral,
even if being gay were a matter of genetics, not choice.
Homosexual sex would still be a violation of the natural
order and gay households would still be a challenge to the
Strict Father family, which is the basis for conservative morality. Gay men are "deviant"; they deviate from the sexual
norms of the community, going outside of the bounds set by
Strict Father morality. Not only are gays seen as immoral in

themselves, but they are seen as a threat since they could lead others "astray," either directly or indirectly through the very existence of homosexual sex and gay households and families, which "blur the boundaries" of moral and immoral behavior.

Perhaps the institution that gays "threaten" the most is the military. President Clinton's proposal, at the beginning of his administration, to allow gays to function openly in the military was attacked violently by conservatives both inside and outside the military. The military is, to a large extent, the institutional realization of Strict Father morality. It has hierarchy, strict roles, punishments and rewards, and requires both physical and moral strength. Discipline is what the army is all about. Though it has a kind of socialistic internal structure (top-down bureaucratic control, government-paid medical care, government-provided housing and schools, PX discounts for members only, no free enterprise, government-provided golf courses and athletic facilities), it serves in the defense of capitalism and it has a Strict Father culture. It has a macho masculine culture. Gay men, despite the popularity of bodybuilding in the gay community, are conceptualized as being weak and feminine, rather than properly macho. Gay men in uniform threaten the image of the uniform: that anyone wearing it is a real man! The masculinity implicit in the meaning of the uniform is anything but trivial. But what really makes gays anathema to the military is all of the ways in which homosexuality flies in the face of Strict Father morality, which is the basis of military culture.

The Clinton administration approached the question of gays in the military as one of civil rights, as if the integration of gays into the military was akin to the integration of blacks and women into the military. It was a drastic mistake. The attempt to integrate gays into the military was seen as an affront to manhood and Strict Father morality all over the country.

Multiculturalism

What does Strict Father morality say about other moral and cultural systems? It says they are immoral. If they do not give primacy to moral strength, they promote moral weakness, which is a form of immorality. They are also immoral if they blur the strict moral boundaries of Strict Father morality, or challenge the moral authority of Strict Father morality, or challenge the Moral Order of our society.

For this reason, conservatives tend to be against multiculturalism, which seeks tolerance for cultural diversity and many other forms of morality. To conservatives, forms of morality other than their own are not moral and therefore not to be tolerated.

Nurturant Parent morality, on the other hand, has a very different view of diversity. Since a nurturant parent gives equal priority to all his or her children, and since children necessarily have differences among them, all those differences have to be respected and toleration is required. Moreover, each child has something different to contribute to the family. Applying the Nation As Family metaphor, diversity in a nation is positive and toleration is required.

Education

Strict Father morality comes with a principle of self-defense: it is the highest moral calling to defend the moral system itself from attack. The very first category of conservative moral action includes acts of promoting and defending conservative morality. The word ''war'' in ''the cultural war'' is not incidental. Conservatives have, at least since the 60s, seen their system of values under attack—from feminism, the gay rights movement, the ecological movement, the sexual revolution, multiculturalism, and many more manifestations of Nurturant Parent morality. Conservatives have seen the values of these movements taught in the schools. They are appalled that what they see as the only system of real morality is being undermined. Conservatives believe that all of the major ills of our present society come from a failure to abide by their moral system. Moreover, they believe that their moral system is the only true American moral system, as well as the only moral system behind Western civilization. They see both of these beliefs challenged in contemporary historical research, which is being taught in our universities. This gives them a sense of moral outrage. They are fighting back.

Why are conservatives in favor of the elimination of the Department of Education, the destruction of the National Endowment for the Humanities, school vouchers, and the privatization of education? Why are liberals opposed to these measures?

A great deal of government support for education is in the form of social programs to help disadvantaged students, and an important charge of the Department of Education has been to develop such programs. Since conservatives see social programs in general as immoral, a quick way to get rid of all social programs in education is simply to abolish the Department of Education and stop funding the programs.

The best research in the humanities these days is not by any means governed by the moral agenda of conservative politics. Indeed, much of it concerns topics that explicitly collide with Strict Father morality and the politics that comes out of it. There is research on ecology, on the heritage of minority groups, on worthy but neglected minority and female figures in history, on the role of American corporations in the exploitation of third-world countries, on the history of unions, on unspeakable things done by the American government and by figures formerly considered heroes, on the value systems of other cultures, on the history of homosexuality, on spousal abuse, and much much more that does not sit well with Strict Father morality and the politics that flows out of it. Getting rid of the National Endowment for the Humanities will eliminate a major source of funding for research by the nation's leading scholars, research that is uncomfortable to conservatives. Meanwhile, private conservative think tanks are funding research that fits a conservative moral and political agenda and the conservative writing of history.

National educational standards are also set by the Department of Education. These standards include things that conservatives would rather not have taught and do not include things that conservatives do want to have taught, such as the recently developed new history curriculum which sets national standards for the teaching of history. Because conservatives have been most effective in changing education at the local level, the elimination of national standards and the leaving of content to local school boards would make it much easier for conservatives to change the curricula in the direction of conservative morality and politics. In other words, the issue seems to be not whether the standards are national or local, but whether they accord with Strict Father morality. Since the promotion of Strict Father morality itself has the highest of values in that moral system, it should follow that conservatives would be happy to have national standards that upheld Strict Father morality.

The privatization of education means that conservatives can set up their own schools in which their children will not have to learn about anything that might be inconsistent with conservative morality and politics. It would also mean a move away from the integration of schools, which means that the children of conservatives would not have to encounter students from different subcultures with different values. School vouchers would make privatization that much easier.

In short, the conservative educational agenda is very much in support of a conservative moral agenda and the politics that it leads to.

From the perspective of Nurturant Parent morality, the issue of education looks very different. Multiculturalism, feminism, gay rights, and the ecology movement are seen, like the civil rights movement, as being great advances—moral advances—in American culture and civilization. Like the civil rights movement, they should be taught as advances. And doing so requires teaching the history of what made them advances, that is, the history of past abuses sanctioned and abetted by major forces in American society and by the government itself. That is not putting down America. On the contrary, it is part of the glory of America that the truth of past abuses by our government and our society at large can be told and the abuses corrected for future generations. Indeed, from the perspective of Nurturant Parent morality, America is a place that has nurtured generations of immigrants. Much of the history of progress in America is the history of what has been made possible through the progressive extension of nurturant morality: progress in equal treatment, progress in opportunities for education and other forms of self-development, progress in health care, progress in humane working conditions, progress in the development of knowledge, and so on. There is a dark side of American history from this perspective as well, and it too must be told: the mass murder of Native Americans and the near extinction of their culture, slavery, the brutalization of factory workers, and the discrimination against women, nonwhites, Jews, immigrants, and gays.

But advocates of Strict Father morality do not see all these changes as advances; they see many of them as immoral, backward steps. And they see the history of some of these changes, which is written from the perspective of correcting abuses, as an attack on their most fundamental moral beliefs. Conservatives are furious at the entire institutional structure of American education. Who runs it? Who gets into education as a profession? Not surprisingly, a great many educators are nurturers. And nurturers often have a Nurturant Parent morality.

After all, teaching small children is not a profession where you make a lot of money. Teachers, with rare exceptions, are not entrepreneurs. Those people who are entrepreneurs and want to live their lives seeking their self-interest through unrestricted free enterprise tend not to choose to teach third grade. Many elementary school teachers are women, often nurturant mothers, so nurturant they want to nurture other people's children.

That is why conservatives are attacking the infrastructure of public education in the country. They have no choice. They are up against an infrastructure full of nurturers, and they don't like it one bit—and they shouldn't like it one bit. They do, however, have allies and a plan of action. Conservative Christians, concerned that public schools teach their children immoral ideas, have been involved in "home-schooling" for years. In addition, many parents who want to insulate their children from such ideas and can afford it have been setting up private schools. But they feel they should not have to pay for public education they don't use—and can't, if they are to control what their children are taught and who they associate with. Such parents have fought for a school voucher system, where government funds for education are distributed to parents in the form of vouchers, which can be used either for private or public schooling. Such a system, if put together cleverly, could destroy much of public education in America. Conservatives would not shed a tear.

THE ISSUE OF STANDARDS

It is generally recognized that a great many American children are not becoming educated. A considerable number of reasons are offered. Educators have pointed to social problems that schools are ill-suited to deal with—drugs, violence, and subcultures where education is not in itself a value. Educators also point to a lack of public willingness to pay taxes to fund education. California, which used to have the best educational system in the nation, has, since the conservative tax revolt of Proposition 13, dropped to near the bottom in per capita spending and has correspondingly suffered a massive loss in quality education.

Conservatives argue that the social problems that afflict schools are the result of permissive parenting and liberal social policies and can be solved by adopting a Strict Father model of the family, conservative political policies, and using private enterprise and competition to produce high-quality schools. Conservatives also argue that liberal approaches to education have lowered educational standards. They talk about "standards" as a hallmark of conservative thought.

Strict Father morality is very much about standards. The metaphors of Moral Authority and Moral Boundaries require absolute standards, imposed by a legitimate authority. Moral Strength and Moral Self-Interest require self-discipline and work, enforced by a system of reward and punishment.

This is true not only of moral standards but of educational standards. The conservative recipe for a good educational system is simply to apply conservative principles and the conservative notion of standards. Teach conservative morality and the conservative notion of character, starting with self-discipline; this is called traditional morality. Set standards based on the classics of Western culture that are tried and true and have withstood the test of time. Make students work hard. Use a system of rewards and punishments: grade seriously and rigorously and fail people who deserve to fail. If there is going to be an elite, it should be an elite of talent, hard work, and achievement. Let those factors determine success. If students fail, they have to take responsibility for their failure, and either do better next time or go through life as failures.

Nurturant Parent morality also carries with it the notion of standards. You cannot grow up to be a good nurturer unless you are disciplined, work hard, and know all that you need to know. Discipline grows out of being raised in a nurturant environment in which you grow up with a responsibility to be empathetic and caring to those around you. Discipline is a no-nonsense matter: people you care about depend on you. There are standards of care and you must learn to meet them. Doing that is difficult every step of the way. You have to work at carrying out your responsibilities, and when others depend on you, you have to be disciplined enough to meet your responsibilities. Learning to cooperate and work with others and meet their standards is essential, and socialization is crucial to this. One of the goals of a nurturance-based moral system is the maximal self-development of each person, partly so that each person can best serve his community and partially so that each individual can feel the deep satisfaction that comes from doing well something that is important to him. Developing one's talents as far as they can go is not easy. It takes hard work and discipline and meeting standards of knowledge and skill.

As with Strict Parent morality, the Nurturant Parent notion

of moral standards also applies to educational standards, which are just as real in both traditions. The difference is not in the standards. The difference is in the very concept of what education to meet those standards should be. Should it be competitive or cooperative? Should the goal be to make students question their teachers and learn through questioning? Or should students just learn to spout fixed answers? Will students learn via the system of rewards and punishments provided by strict grading? Or will students learn because they are interested, because it matters to them, and because they want to please their teacher and to function as well-liked, respected, and responsible members of the class? And what shall we do with less talented students? Should they just be left to fail and drop out? Or is it better to find a way to keep them in school and let them learn as much as they are able? The sink-or-swim approach of Strict Father morality says let them fail and drop out. The Nurturant Parent approach of maximizing self-development says find a way to let them learn as much as they can.

From the point of view of the Nurturant Parent approach, the issue of standards is a red herring. Everyone sets standards. The question is what standards to have, what should be done in the classroom to meet them, and what education is about. The two approaches say very different things.

In the Nurturant Parent approach to education, the ability to nurture successfully requires honest inquiry; we must know ourselves and our history, not only the bright side, but more importantly, we must know the dark side of America. We must learn the dark side partly so that we will not repeat it, partly so that we can appreciate progress and what is good about America, and partly so that we will not be self-righteous. A truly nurturant education is not a goody-goody education. It involves teaching honestly about controversial and socially volatile issues, and understanding why they are socially volatile. It does not distort either the past or the present to make us feel good. It tells the truth. It encourages questioning. It is not just a matter of learning facts. It involves understanding what those facts mean in a larger context—both a historical context and a contemporary one. It also requires understanding differences in points of view, in learning that there is not just one way to view events.

Art

Strict Father morality has major implications for the conception of art. Art can be seen as having a value if it serves a moral purpose, say, to build moral strength and character or to display moral modes of life. Let us call this morally correct art.

In Strict Father morality, art can also be something beautiful, appreciated for its craftsmanship, or otherwise enjoyable or uplifting. From this perspective, art is a form of wholesome, uplifting entertainment, and many newspapers and magazines include reports of art in their entertainment sections. As entertainment, art is not in itself a serious enterprise of social importance—it doesn't go in the public affairs section of the newspaper or in the section with science or business.

Art, as a source of pleasure or entertainment, also makes sense as a commodity, something produced that plays a role in the economic system. Seeing art as a commodity makes the prices commanded by works of art newsworthy, and we constantly see news stories about how much a work of art has been sold for.

Moreover, good art, as a source of pleasure and/or inspiration as well as a solid investment, becomes a symbol of success. As a source of pleasure, it is a reward for hard work. And as a good investment, it is a measure of your ability to make such an investment. That is why owning an original artwork is a symbol of success. The value of art, in Strict Father morality, lies either in its moral value, its entertainment value, its economic value, or its value as a success symbol—a sign of belonging to an elite.

All these views of art impose on art a set of standards—moral standards; standards of beauty, craft, or entertainment value; and standards for lasting economic investment. And since Strict Father morality requires that standards be absolute and lasting, it is not surprising to find conservatives applying such views to art.

The best known conservative art journal is the *New Criterion,* edited by Hilton Kramer, former art editor of the *New York Times.* The word "criterion" was chosen carefully. The purpose of the journal is to impose and uphold standards for "worthwhile" art. It also has the purpose of criticizing what it sees as immoral, unbeautiful, unskillful, unentertaining, or of no lasting value. The best art is seen as moral, beautiful, uplifting, and lasting.

Nurturant Parent morality also motivates various views of art. As a moral system, it approves of art with a moral purpose, art that serves Nurturant Parent morality, art with the right kind of social message. In this way, it is like Strict Father morality, only the message is different. It has another form of morally correct art, art that looks at the moral system from the outside and says, "This is the correct moral system to have."

But there is a difference between art that is about the choice of a moral system and art that is motivated by the internal workings of that moral system—art that is about the choice of nurturance and art that *is* nurturant, in one of a number of ways. The first is art with a liberal message. The second is art within a liberal artistic tradition.

The idea of morality as happiness gives rise not just to art that is beautiful and pleasurable, but also to art that is playful and fun.

The liberal tradition in art, since it derives ultimately from nurturance, promotes the kind of artistic thought that true nurturance requires: it promotes art that is questioning and probing, that forces one to look at the dark side of life and won't allow denial, that forces one to face unpleasant truths and leads one to undergo self-exploration and reevaluation.

In addition, since Nurturant Parent morality promotes self-nurturance and self-development, it supports the notion of art as a meditative experience, a development of one's imaginative capacity, an exploration of perception, and an exploration of form and of materials.

Considerations of empathy, which are central to nurturance, lead to an art of exploring the other—other cultures, other subcultures, unusual people or places or events. Part of this impulse is multicultural art, which places us into another worldview. The application of nurturance to art also leads naturally to art as a form of healing, perhaps as a self-expressive therapy for the artist, perhaps as a form of healing for the culture.

Finally, some artists, reacting to the commodification of art that they identify with the conservative tradition, prefer art that is de-commodified—nonobject art, such as conceptual art or performance art. The idea here is that art is a matter of experience and need not be a commodity.

Much of contemporary art falls under one of these categories, and can be seen as deriving from some aspect of Nurturant Parent morality. From the perspective of Nurturant Parent morality, such forms of art are moral enterprises. If liberalism is the application of Nurturant Parent morality to politics, then there is a good reason why most artists in these traditions are political liberals.

Since education about contemporary art is abysmal in this country, most citizens—well-educated liberals included—know very little about many forms of the art discussed above. Moreover, tastes differ. Some liberals like morally correct liberal art and find it uplifting; others hate it as being crass.

Some simply like art to be beautiful or pleasurable. But only a small minority of liberals are educated as to most of the forms of contemporary art listed above. As a result, most liberals do not know or care enough to defend it from conservative attack.

And conservatives do attack art in this tradition. Highbrow attacks come from Hilton Kramer and his colleagues at the *New Criterion*. Political attacks come from Jesse Helms and others in Congress who want to destroy the National Endowment for the Arts. Conservatives, not surprisingly, hate morally correct liberal art and prefer morally correct conservative art. That is, they hate art promoting feminism, gay rights, multiculturalism, government programs, and so on. Robert Mapplethorpe's beautiful, moving, and sometimes disturbing photographs of gay men outraged conservatives. Conservatives also hate art that probes the dark side of American life and American history and forces us to confront the not-very-nice facts about our country, as well as art concerning deep conflicts within our culture. André Serrano's crucifix immersed in urine is a physical metaphor for a Catholic's rage at his own church. For those who see art from the perspective of Nurturant Parent morality (and not all liberals do), this piece raised questions about the dark side of the church's role, not only in the artist's life, but also in the lives of its communicants and in society in general. But to conservatives, for whom raising such questions is not a legitimate function of art, Serrano's work was seen as no more than an insult to religion. Other works—say, works of conceptual art or explorations of form—are incomprehensible to the conservative value system. Conservatives only express puzzlement and cannot see why they are art at all, rather than silliness or self-indulgence. Indeed, from the perspective of Strict Father morality, they are not art. The only option open to conservatives is dismissal—to see such art as a product of a "cultural elite"—"effete snobs" whose values are illegitimate. It is thus no surprise that conservatives want to eliminate the National Endowment for the Arts.

The term ''cultural elite'' must also be understood in terms of Strict Father morality. The word ''elite'' has one sense which indicates superiority and prestige, often with the connotation that the prestige is undeserved and the superiority illusory; it also has the sense of a prestigious, self-sustaining in-group. Strict Father morality subordinates culture to morality, its own morality. The idea of a real cultural superiority that isn't moral superiority makes no sense from their perspective. For this reason, the term ''cultural elite'' can only be ironic, referring to a self-sustaining influential group with false claims to superiority. The implication of the very use of the term ''cultural elite'' is that the truly superior people are those who are morally superior, namely, those who abide by Strict Father morality. They are the true elite who should be culturally celebrated, while the so-called ''cultural elite,'' who do not place Strict Father morality above other cultural values, are immoral and should be brought down.

Art in America is seen from a moral perspective. It can be seen as serving a moral system, as in morally correct art, or it can be seen as serving some function within a moral system—a pleasurable reward, a commodity, a sign of success, a form of questioning, an exploration of perception, a mode of healing.

Because of the centrality of art in our very notion of culture, art is bound to be a battleground in the culture war, a war that conservatives are pursuing vigorously, and a war which they perceive, perhaps accurately, as one of self-defense.

Moral Education

Conservatives have understood very well that politics is based on morality. They have also understood intuitively that Strict Father morality underlies conservative political thought. To spread their views to the coming generations, they are doing exactly what they should be doing: working to install Strict Father morality as the official version of morality in the schools. Here the main figure is William Bennett and *The Book of Virtues* is his primer.

What Bennett calls "traditional moral values" are the values of Strict Father morality. Though Nurturant Parent morality may have been traditional for mothers throughout history, it has gained public prominence in popular American culture mainly within this century as women have gained more prominence and influence. Because of the cultural dominance of strict fathers over the centuries, the tradition of Nurturant Parent morality has been less visible, but it is no less "traditional."

It should be clear by now that teaching Strict Father morality—so-called "traditional" morality—is teaching political conservatism, which is why a political figure such as Bennett has put so much effort into it. To be in charge of the teaching of morality is to be in charge of the teaching of politics.

The Book of Virtues is interesting for the virtues *not* taught there: nurturance, tolerance of "deviants," social responsibility, open-mindedness, self-questioning, egalitarianism, championing the disenfranchised, communion with the natural world, aesthetics for its own sake, self-development, being in touch with one's body, or happiness (in the sense of Moral Happiness). There are no sections on modern moral issues: the equality and independence of women, the morality of organized labor, or the protection of consumers and the environment. These are other virtues very much worth teaching—especially to children. They are among the virtues of Nurturant Parent morality.

Is there a way to provide moral education without engaging in political indoctrination? Fortunately, there is. It would be the teaching of how morality is understood in this culture, both Strict Father and Nurturant Parent forms of morality and the critiques each makes of the other.

Here are some guidelines. Teach the two models of the family, and the two moral systems that fit them best. Relate these family-based moral systems to issues of gender. Teach what Strict Father morality is and what Nurturant Parent morality is, and what the differences are. Point out that they don't agree and that this is a political as well as a moral matter. Give examples. Point out how each would criticize the other. When you teach Strict Father morality, by all means exemplify Bennett's list of virtues. But when you teach Nurturant Parent morality, exemplify all the virtues listed above that Bennett does not include. Be sure students understand which virtues fit which view of morality. And be sure they understand that this is a political as well as a moral matter. So far as I can see, this is the only way to keep the teaching of morality from being political indoctrination.

There may be a problem in taking this route to teaching morality and avoiding partisan moral and political indoctrination. Many conservatives believe that there is only one possible view of morality—Strict Father morality. Many religious conservatives believe that teaching both moral systems is itself immoral. Because it promotes discussion about what morality is, it prompts children to think for themselves rather than merely obeying authority. If one follows my suggestion, some may argue that even equal time for Nurturant Parent moral views subverts traditional morality just because it teaches children to think for themselves and not just obey authority.

It is important to resist such arguments. Children need to be taught moral ideas, which includes the fact that there are alternative moral views that are deeply grounded in the most basic of human experiences, namely, family life. They need to know that there are alternative views of what family life and family-based morality ought to be. They need to know that it is not just a matter of flipping a coin, of saying one moral system is just as good as another. There are facts that matter, which will be discussed below in Chapters 20–23.

Finally, it is crucial that the teaching of morality not be left only to religion. Each interpretation of a religious tradition chooses a form of morality to go with it. It is important that children have an education, outside of their particular religious tradition, that allows them to understand what form of morality is built into their religious teachings. It is important for children, as they are growing up, to be able to see morality from the outside as well as the inside, to be able to name the features of a moral system and to see the differences across systems. In a democratic country we have to live with people who have very different moral views than our own. We need to be able to pick out those views, name them, and discuss them openly.

The issue of moral education is an explosive issue. It raises the deepest questions of who we are and what we should raise our children to be. There are no more important questions than these.

Summary

Strict Father and Nurturant Parent moralities come into play across a whole spectrum of American life. In affirmative action, they show up in the distinction between the Strict Father focus on head-to-head competition versus the Nurturant Parent focus on the overall well-being of the family (metaphorically, the nation) over whole groups and long periods of time, taking into account such practical matters as subculture differences and the effect of cultural stereotyping on individual judgments.

The division manifests itself in the issue of gay rights as a matter of upholding the required heterosexual orientation of the Strict Father model over the equal nurturance ethic of the Nurturant Parent model. In the multiculturalism issue it is a matter of upholding a single standard of authority versus the equal nurturance ethic. In education, it is a matter of upholding the Strict Father moral system itself versus the internal questioning and need for understanding required for true nurturance. In art, family-based moral views provide very different conceptions of what art should be. And in the teaching of morality, the issue is whether Strict Father morality is to be taught as the only morality, or whether a discussion of these two major alternative moral systems should be held. The culture war shows itself most clearly in the issue of moral education.

Because of the bitterness of the culture war it is necessary to understand what the ultimate sources of division are because they manifest themselves so differently in political issues.

14

Two Models of Christianity

There are those who claim that their politics is simply a matter of following what the Bible literally says. This assumes that there is such a thing as a literal interpretation of the Bible. Indeed, whole branches of Christianity are based on such a claim, but it is straightforwardly a false claim.

Nobody believes that "The Lord is my shepherd" is said literally by a sheep that has fleece and eats grass. Nobody believes that "Our Father Who art in heaven" is literally daddy. Virtually every page of the Bible is filled with passages that can only be, and always are, interpreted metaphorically. There simply is no fully literal interpretation of the Bible.

There is nothing surprising about this or wrong with it. Literal modes of thought and literal language are simply not adequate for characterizing God and the relation of human beings to God. Such things can be understood only through metaphorical thought and communicated through metaphorical language. God, after all, is ineffable—beyond human comprehension. If you're going to even think or talk about God, you're going to have to use human experience as your basis and have an extensive collection of metaphors at your disposal. My colleague, Professor Eve Sweetser, has been studying metaphors for God in the Judeo-Christian tradition. She began, somewhat arbitrarily, by looking at the Judaic tradition, at the metaphors in the Yom Kipper liturgy, at a point where a list is given. The list is a good starting point; it provides a representative collection of Judeo-Christian metaphors for God, since many of the Christian metaphors were inherited from the Jewish tradition. Here is the list:

God is a father; humans (or specifically Jews) are his children.
God is a king; human beings are his subjects.
God is a male lover; humanity (or the Jewish people) is his female lover.
God is a shepherd; humans are his flock of sheep.
God is a vineyard-keeper; humans are his vineyard.
God is a watchman; humans are the treasure he guards.
God is a potter; humans are his clay.
God is a glassblower; humans are his glass.
God is a smith; humans are his metal.
God is a helmsman; humans are the rudder (or ship).
God has chosen us; we have chosen God.

As Sweetser (in a work in preparation) points out, this collection of metaphors for God forms a radial category, with God As Father at the center. The God As Father metaphor is the only metaphor that overlaps in one way or another with each of the other metaphors.

- The father and king metaphors both attribute *authority* to God.
- The father and lover metaphors both attribute *nurturance* to God and posit *mutual love* between God and human beings.
- The father, king, shepherd, and watchman metaphors all attribute *protectiveness* to God.

- The father, vineyard-keeper, potter, glassblower, and smith metaphors all attribute to God a *causal ontological relationship:* bringing people into being.
- The father, lover, and choice metaphors all see the relationship as between *two volitional beings.*

These metaphors are not to be seen as operating simultaneously. For example, God is not to be seen as a father who is also a male lover; that would make the relationship incestuous, and it is certainly not intended to be that. The metaphors, though overlapping in the understanding of God that each posits, are not to be seen as holding all at once.

This collection of metaphors for God should give some sense of the richness of metaphorical thought in religion. Since the Bible cannot be understood literally, it must be interpreted. Different denominations of Judaism and Christianity accept different interpretations of the Bible, each denomination, of course, accepting only its own interpretation. It should be clear, then, that the claims of certain Christians that their politics is just a matter of following what is literally in the Bible must be false, though I do not doubt their sincerity in believing that claim.

This raises important questions: What makes conservative Christians conservative? What makes liberal Christians liberal? Christianity in itself takes many political forms, from the conservative Christian Coalition to the liberal Interfaith Alliance, the National Council of Churches, and even liberation theology in Latin America and elsewhere. There are over 400 Protestant denominations, and it appears that a majority are liberal and a well-organized minority are conservative.

WHAT MAKES CONSERVATIVE CHRISTIANS CONSERVATIVE?

My guess is that what makes conservative Christians conservative is that they interpret their religion as requiring a Strict

Father model of the family and Strict Father morality. The God As Father metaphor attributes to God both authority and nurturance. But there are various possibilities for how authority and nurturance can go together. In the Nurturant Parent model, the child's obedience to the parents' authority is a consequence of the parents' proper nurturance. In the Strict Father model, the reverse is true: authority comes first. First and foremost, the child must obey and not challenge the strict father's authority; to the obedient child, nurturance then comes as a proper reward (see References, B3; Dobson, 1992, pp. 20–22).

One's relation to God can be interpreted in either way. On the Nurturant Parent interpretation, you accept God's authority because of his original and continuing nurturance. On the Strict Father interpretation, God is seen as setting the rules and demanding authority; if you obey, you get nurturance. The difference is one of priorities, and, as we have seen, that is an all-important difference.

To place that difference in perspective, I will have to give a schematic account of certain aspects of conservative Christianity, using the metaphor system discussed throughout this work. To do this without anything extraneous clouding the picture, I will have to give the barest of bare skeletons of the religion. I ask your indulgence for my oversimplification, which will of necessity sound like the text of a comic book called, "Christ for Beginners."

I, of course, assume that there are a great many variations on this model, just as there are variations on Strict Father morality and conservative politics (see Chapter 17). Moreover, I realize that one could have a Nurturant Parent model of the family and a Strict Father model of religion, or the converse. What I am describing here are the maximally coherent ideologically pure positions.

Christianity in general associates the nurturant aspect of God with Christ. Christianity works by a moral accounting system inherited from Judaism. Immoral deeds are debits; moral deeds are credits. By moral arithmetic (see Chapter 4), receiving something of negative value (say, suffering) is equivalent to giving away something of positive value (as in paying for something). Suffering is thus a way of paying for your sins: suffering builds up moral credit, just as doing good does. If you have a big enough positive balance of moral credit when you die, you go to heaven; if you have a negative balance, you go to hell. These general notions are shared by most forms of Christianity. At this point, let's look specifically at Strict Father Christianity.

Strict Father Christianity

Being made of flesh, human beings are morally weak. This inherent moral weakness is called Original Sin, as exemplified by the moral weakness of Adam and Eve, which resulted in God's taking everlasting life away from human beings. Because of their moral weakness, everybody starts off with a large moral debit—big enough to guarantee that ordinary people would go to hell.

But God loved human beings so much, he wanted to offer them a way out of this horrible fate which arose from their inherently sinful bodily nature. So he made his only Son a human being who was free of sin, and hence had no moral debits. Then God allowed his Son to be crucified, and in so doing had him suffer more than the total possible suffering of all mankind forever. Through all this suffering, Jesus built up a huge amount of moral credit, much more than enough to pay for the Original Sin of all mankind. Through his crucifixion, Jesus paid off mankind's Original Sin debit. This made it possible for human beings to go to heaven if they were righteous enough.

But there were those human beings who, over the course of their lives, had sinned so much and run up such a moral debit that they were destined for hell no matter what they did for the rest of their lives. But Jesus loved all people so much, including those sinners, that he suffered on the cross enough to pay off their moral debts too. This was an enormous act of love, but not unconditional love. There had to be a condition. It would have been wrong to let wrongdoers get into heaven without doing anything at all to get there. That would have undone the whole point of the moral accounting system, which is to get people to follow God's commandments.

So Jesus offered sinners a deal. If they would truly repent, accept him as their Lord, join his church, and follow his teachings for the rest of their lives, he would pay off their moral debts with the moral credit from his crucifixion and wipe their slate clean. It would be as if they were born again, with no moral debits. That way he would save them from hell; he would be their Savior. The contract was made available to all sinners at any time.

As their part of the deal, the former sinners would have to accept the authority of God and follow his commandments for the rest of their lives. This would be hard. It would require a character one did not have before being born again, a new moral essence—not being rotten to the core, but being rock solid.

To acquire this moral essence, you have to take Jesus into your heart. The heart is the metaphorical locus of moral essence. You have to take the essence of Jesus into you and make it your essence. That is not as easy as it may sound.

It requires building up moral strength through self-discipline and self-denial. It requires obeying moral authority, the moral authority of God, as revealed through the Bible and his church. It requires staying within moral boundaries and not deviating from the path of righteousness. And it requires remaining pure and upright. Jesus' offer was one of love—not unconditional love, but tough love.

Unfortunately, there is a large loophole in this contract. Allowing people to repent at any time and still go to heaven is an incentive to keep sinning as long as possible. Then, at the last minute, you can repent and go to heaven. That's not an incentive to be good; it makes sinning pay off up until you're about to die.

This loophole is closed by the Last Judgment, the idea that, at some time which human beings cannot predict, the world will come to an end and the moral books will be closed. At that instant, you will be judged, and if you are a sinner—if you have more debits than credits at that instant—you will be forever damned to hell. Because the Last Judgment could happen at any time, the only way to guarantee that you will be able to take advantage of the deal Jesus has offered is to accept it immediately. If you truly repent now, you guarantee yourself a spot in heaven. On Judgment Day, Jesus will be your Redeemer: He will pay off your old moral debts and redeem you from the clutches of the Devil.

PROPERTIES OF THE STRICT FATHER INTERPRETATION OF CHRISTIANITY

What makes this biblical interpretation a Strict Father interpretation is the centrality of God's authority, his strict commandments, the requirement of obedience, the priority of moral strength, the need for self-discipline and self-denial, and the enforcement through reward and punishment.

What makes this system work is reward and punishment based on moral accounting—reward for following, and punishment for not following, the authority of God. Without it, the entire moral system makes no sense. Without the reward of heaven, people will not follow God's commandments and all morality will break down. An offer of redemption without making you work for it would be like welfare: it would be an incentive to be immoral.

It is this religious system of reward and punishment that James Dobson, in the quotation above, claimed (through metaphor) to be the same as laissez-faire free-market capitalism. What conservative Christians have done is two things. They have given the Bible an interpretation in terms of Strict Father morality and they have, through metaphor, linked that interpretation of the Bible with conservative politics.

The Bible can, of course, be interpreted in many ways—and the wildly different denominations of Jews and Christians testify to that. For example, when a sinner accepts Christ as his Savior, he could see himself as appreciating Christ's, and hence God's, nurturance and thereafter taking Christ as his model for nurturant behavior. The former sinner would thereafter adopt a Nurturant Parent morality. If, in interpreting the Bible, one gives priority to God's compassion and nurturance rather than to his authority, then one's religious morality will tend to be Nurturant Parent morality and the application of that morality to political life would tend to make one a liberal. We will discuss such an interpretation shortly.

It is important to understand just why conservative Christians are conservative, and to realize that their interpretation of the Bible gives them no special claim to morality. Indeed, in this case, family-based morality *precedes* morality based on religion, since it is Strict Father morality that yields this interpretation of the Bible.

Conservative Christians are not conservative because they interpret the Bible—all of it—purely literally. There may be parts certain denominations do take literally, such as that God created the world in exactly 168 hours. But in many of the most important respects, conservative Christians have a metaphorical interpretation of the Bible just like everyone else. What makes them conservative is the same thing that makes them the kinds of Christians they are—the use of Strict Father morality. Indeed, that's what makes conservative Christians try to interpret the Bible literally. If the Bible, as the word of God, contains commandments from the highest moral authority, commandments which are to be followed absolutely and strictly, then someone seeking to strictly obey those commandments is committed to having no personal, subjective interpretations; he would not be following *God's* commandments if they were only *his* interpretations of God's commandments.

But as we have seen, this is a self-defeating enterprise. Purely literal interpretation of the Bible is impossible—even in its most important respects—the understanding of God, one's relationship to Him, and the overall understanding of the conservative Christian tradition. In conservative Christianity, all of these require an interpretation—a Strict Father interpretation.

But conservative Christians go beyond applying Strict Father morality just to religion. They apply it to politics as well, forging a metaphorical link between (1) their religious system of moral accounting, (2) laissez-faire free-market economics, and (3) the Strict Father morality system of reward and punishment.

What makes conservative Christians conservative is this system of metaphor: God As Father, Well-Being As Wealth and Moral Accounting, Morality As Strength, Morality As Obedience, Moral Self-Interest, and so on. The Morality of Reward and Punishment, for example, assumes the metaphor of Moral Accounting and would make no sense without it.

In conservative Christianity, laissez-faire free-market capitalism is a moral, not just an economic, enterprise. Why? What makes it moral, not just economic? After all, the economics just says that, if each person individually pursues his own financial profit, the financial profit of all will be maximized. To make this a thesis about morality, you have to conceptually project financial profit onto moral profit. The metaphor by which this is done is Well-being As Wealth.

Only then do we get a moral thesis, which I have called Moral Self-Interest: If each person individually pursues his own self-interest, then the self-interest of all will be maximized.

Interestingly enough, the economic theory is itself based on a metaphor, Adam Smith's metaphor of the Invisible Hand. The Invisible Hand metaphor makes use of the notion of force and the very general metaphor that Control Is the Exercise of Force by the Hand, as in expressions like "It's in your hands now," "They handed him over to the FBI," "It's out of my hands," "I can't handle this project," "You're in good hands with Allstate," and so on. The metaphor posits invisible economic "forces" that control the operation of a free market. In this theory, economic "forces" are understood metaphorically in terms of physical forces.

The link in conservative Christianity between laissez-faire free-market capitalism and morality is thus metaphorical through and through, as is the interpretation of the Bible via Strict Father morality, and Strict Father morality itself is metaphorical through and through. To say this is not to "deconstruct" and question the validity of Strict Father morality and conservative Christianity, it is merely to apply an analysis of the concepts involved to enterprises of Biblical interpretation. I cannot repeat often enough that there is nothing wrong with metaphorical thought in itself. No one can think or function without it.

CAN YOU BE A BORN-AGAIN CHRISTIAN AND NOT A CONSERVATIVE?

Of course you can. First, you could have a Nurturant Parent interpretation of the Bible. Second, if you have a Strict Father interpretation, you need not use the Nation As Family metaphor to project Strict Father morality onto the political domain. That is, you could keep your religion private. Third, you need not use the metaphorical link between the Christian system of moral accounting and laissez-faire free-market economics.

Conservatism and born-again Christianity are distinct systems of thought. Conservative Christianity links them together by metaphor. That link need not exist. It is a matter of interpretation.

Nurturant Parent Christianity

To put the Strict Father interpretation of Christianity in perspective, let us look at the Nurturant Parent interpretation. The fundamental metaphors used in the interpretation are these:

- God Is a Nurturant Parent to Human Beings.
- Christ Is the Bearer of God's Nurturance to Human Beings.
- God's Grace Is Nurturance.
- Moral Action Is Nurturant Action (helping through feeling empathy, demonstrating compassion, acting out of love, etc.).
- Immoral Action (Sin) Is Nonnurturant Action toward Others (harming through lack of empathy, lack of compassionate action, etc.).

God's grace is the central notion in this interpretation, and it is understood metaphorically as nurturance. Since nurturance is a rich concept, the Grace As Nurturance metaphor leads to a correspondingly rich notion of grace. Here are some of the basic properties of nurturance:

PROPERTIES OF NURTURANCE

Nurturance is the expression of a parent's Love.
Nurturance entails the Presence of the parent.

Nurturance entails the Closeness of the parent.

Nurturance involves Feeding that allows for Growth.

Nurturance involves Healing.

Nurturance results in Happiness.

Nurturance entails Protection.

Nurturance is not earned.

A nurturant parent gives nurturance freely and unconditionally.

Nurturance must be accepted if a child is to receive the benefits of nurturance. Only through receiving nurturance do children learn to be nurturant to others (to feel love, feel empathy, act for the benefit of others, and so on).

To learn to act nurturantly toward others, one must be nurtured by a nurturant parent.

The Grace As Nurturance metaphor projects these aspects of nurturance onto corresponding properties of grace. The result is a metaphorically rich notion of grace.

PROPERTIES OF GRACE

God's Grace is the expression of God's Love.

God's Grace entails the Presence of God.

God's Grace entails the Closeness of God.

God's Grace involves a spiritual Feeding (or filling) that allows for moral Growth.

God's Grace Heals spiritual wounds and spiritual ills.

God's Grace results in Happiness.

God's Grace entails God's Protection.

Grace is not earned.

God gives his Grace freely and unconditionally.

God's Grace must be accepted if one is to receive the benefits of Grace.

Only through receiving God's Grace do people learn to act morally toward others (to feel love, feel empathy, act for the benefit of others and so on).

To learn to act morally toward others, one must receive
God's Grace (e.g., one must be in God's presence, be
close to God, etc.).

Since grace is understood metaphorically as nurturance, it is
of course the central concept in Nurturant Parent Christianity. The central need in this version of Christianity is the need
for God's Grace—his nurturance along with his presence and
closeness—if one is to learn to act morally and if one is to
be spiritually filled, sustained, and healed. What gives rise
to this need is a separation from God, a separation that is
natural to the human condition.

Original Sin in the Nurturant Parent interpretation is rather
different than in the Strict Father interpretation, as is sin
itself. Sin is nonnurturant action toward others. The idea
behind Original Sin is that we naturally act nonnurturantly
toward others and we must learn to act nurturantly.

The way we learn to act nurturantly (that is, morally) is
through being nurtured and then imitating the actions of those
who nurture us, thus incorporating the capacity for nurturing
into ourselves. But our own parents are never perfect nurturers, and may be very imperfect ones. Moreover, since we
naturally start separating ourselves from them at an early
age—in toddlerhood—we can never receive full and constant nurturance from our parents. The very best of nurturant
parents cannot come close to making us perfect nurturers,
that is, fully moral beings. Original Sin, then, is our inherent
inability to be fully moral (that is, nurturant) beings for these
reasons. It is also the state of being born and raised separate
from God, the ultimate nurturant parent. The task of becoming a fully nurturant person is thus to find God, the ultimate
nurturant parent, and receive his perfect continuing nurturance—his Grace.

On the Nurturant Parent interpretation of the Bible, Eden
is the state of being completely loved and nurtured in early

infancy. Eating of the fruit of the tree of the knowledge of good and evil is reaching the stage at which we begin to separate from our parents, follow our own desires, and have to learn to act morally toward others. The expulsion from Eden is reaching the stage at which we no longer are completely loved and nurtured, as in infancy, and have to face the difficulties of the world and learn to act morally. Thus we are faced with the problem of finding God and, through his Grace, growing morally and becoming as fully nurturant as we can be.

Let us now turn to how Christ fits into this picture. I begin with the general notion of Moral Accounting, as it applies to both the Nurturant Parent and Strict Father models:

MORAL ACCOUNTING

The Moral Accounting metaphor works as follows:

- Acting immorally gives one a moral debit.
- Acting morally gives one moral credit.
- Suffering (by moral arithmetic) gives one moral credit.

If one's credits exceed one's debits at death, one goes to heaven. If one's debits exceed one's credits at death, one goes to hell.

Heaven is the state of receiving perpetual nurturance from God: being in God's presence, being close to God, being filled with his love and warmth, being blissfully happy. *Hell* is the state of never again receiving nurturance from God: being alienated from his presence, being unloved, uncared for, unhappy.

Because of Original Sin, people inevitably act immorally and run up huge moral debits, so huge they can never pay them off with good deeds. On the basis of our own actions alone, therefore, all human beings deserve to go to hell, to suffer alienation from God and his nurturance in perpetuity.

But God, being a nurturant and loving parent, did not want his children to have to suffer forever. He wanted to give them a chance to go to heaven, to be with him forever and be nurtured forever. But since human beings had earned eternal damnation, their moral debts had to be paid off for them to have a chance to enter heaven. They would have to grow morally and learn to act nurturantly toward others. The only way they could do this would be through example, through the example of a perfectly nurturant human being. Even so, human beings could never do enough good deeds to pay off their moral debts. Those debts could only be paid off through human suffering.

Since God, being God, could not teach human beings nurturance through human example, and since he, not being human, could not suffer human afflictions to pay off the moral debt of human beings, another solution was required. God sired a human son—Christ—who could do these things for the sake of saving human beings from hell.

Christ, being the son of God, was without sin; that is, he acted with complete nurturance—with empathy toward others and for the benefit, development, and happiness of others, requiring nothing in return. Through his actions, he set an example of a perfectly nurturant person for human beings to follow, so that they could learn to become nurturant toward others.

Christ, being perfectly nurturant, made the ultimate sacrifice for human beings. On the cross, Christ suffered so much that his suffering more than balanced the sins of the entire human race. Through his suffering he acquired enough moral credit to pay off the moral debts of all human beings.

Christ thus delivered God's nurturance to human beings, that is, he offered us God's Grace. Through accepting Christ and following his example of nurturance, we can grow morally, gradually incorporating his nurturance into ourselves and becoming nurturant beings.

In those denominations of Christianity where communion is of great importance, communicants are brought at communion into God's presence: they are offered God's Grace and accept it. In eating the wafer and drinking the wine that is symbolic of Christ's body and blood, they are incorporating the essence of Christ—nurturance—into themselves, and hence changing their own essence to become increasingly nurturant.

In accepting Christ, in following his nurturant example, one is accepting God's Grace. Christ, being the ultimate nurturer, offers Grace to all. Those who accept, those who "take Christ into their hearts"—that is, who take nurturance as the essence of their being—have their moral debts paid off with Christ's moral credit. They are "redeemed" and thus are saved from hell and can earn their way into heaven through their subsequent nurturant behavior.

Comparison

The Nurturant Parent interpretation of Christianity has very different consequences than the Strict Father model. First, it presents a completely different view of the proper relationship between human beings and God.

In Strict Parent Christianity, God is a moral authority, and the role of human beings is to obey his strict commandments. The way you learn to obey is by being punished for not obeying and by developing the self-discipline to obey through self-denial.

In Nurturant Parent Christianity, God is a nurturer and the proper relationship to God is to accept his nurturance (Grace) and follow Christ's example of how to act nurturantly to others. There are no strict rules; rather one must develop empathy and learn to act compassionately for the benefit of others, whatever that might require. You learn to become nurturant through receiving nurturance, through accepting the pleasures of nurturance, developing, growing, and following the example of the ultimate nurturer (Christ).

The two forms of Christianity assume very different views of human nature. Strict Father Christianity assumes folk behaviorism, that people function to get rewards and avoid punishments, and that discipline and denial build character. Nurturant Parent Christianity, on the other hand, does not assume folk behaviorism nor the need for discipline and denial to build character. Instead, it assumes that being nurtured builds the right kind of character (nurturant character) and that those who are nurtured will thereby incorporate nurturant instincts into them.

The two forms of Christianity assume different ideas of what a good person is. Strict Father Christianity assumes that a good person is one who is self-disciplined and self-reliant and who can function well in a hierarchy, someone who can obey strict orders from those above and give strict orders to those below—and enforce those orders with pain. Nurturant Parent Christianity sees a good person as one who is nurturant, one who can function well in interdependent situations, where social ties, communication, cooperation, kindness, and trust are essential.

Finally, the two forms of Christianity have very different understandings of what the world should be like so that such ideal persons can be produced. Strict Father Christianity requires that the world be competitive and survival difficult if the right kind of people (strong people) are to be produced and rewarded. Nurturant Parent Christianity requires that the world be as interdependent, nurturant and benign as possible, if the right kind of people—nurturant people—are to be produced.

In short, these two models of Christianity directly reflect the social values of the models of the family on which they are based. What this comparison shows is that there is no neutral Christianity. One can have a Strict Father interpretation or a Nurturant Parent interpretation, or many variants on these, and no doubt many other interpretations. In giving an interpretation of the Bible, one has a choice as to which passages to pay most attention to, or give most weight to, and which passages to ignore or give less weight to.

Thus, one cannot just point to the Bible and say that it, in itself, gives one or the other interpretation. Nor does the Bible, in itself, prefer one or the other moral system or family model. Instead, family models are imposed on the Bible in providing interpretations. Thus, conservative Christians have a conservative interpretation of the Bible because they apply the same Strict Father family model to their Christianity as they do to their politics. Similarly, liberal Christians have a liberal interpretation of the Bible because they, correspondingly, apply the same Nurturant Parent family model to their Christianity as to their politics.

Summary

There is no such thing as a literal interpretation of the Bible, though there are both conservative and liberal Christians who choose passages to interpret as literally as possible and who claim to have a literal interpretation. Nonetheless all interpretations are metaphorical. Strict Father and Nurturant Parent interpretations give rise to conservative and liberal forms of Christianity.

The Bible, in itself and without interpretation, can say nothing at all about the kind of politics that one should have. It is only through Strict Father and Nurturant Parent interpretations of the Bible that one is led to a conservative or liberal religious politics.

Variations and mixed views do, of course, occur; the account given here is intended only to cover the central cases.

15

Abortion

Let us begin with some lexicography: the words *embryo, fetus,* and *baby. Embryo* and *fetus* are medical terms. An *embryo* is a product of conception more organized than just a cluster of cells but not yet recognizable as a member of the species. A *fetus* is a further stage of the organism but one not yet born. There is no precise, objectively specifiable moment at which a *cluster of cells* becomes an *embryo* and the *embryo* a *fetus.* But there are certain relatively clear cases.

Take the cases of an IUD and a morning-after pill. Both of those prevent a fertilized egg that has become a cluster of cells from becoming attached to the mother's uterus. The cells are then expelled and the result looks like spotting from the woman's period. The cluster of cells is not yet an embryo.

The embryo turns into a fetus—a recognizable form of the human species—at some time between eight and twelve weeks. The fetus at this point cannot exist outside the woman's body until, at the very least, several weeks later. An embryo then is not recognizable as a form of the species and cannot have an existence separate from the mother's body. Once the *fetus* is born, it is called a *baby*.

The terms *embryo* and *fetus* call up a medical context, in which the issues are medical issues. An abortion from the medical perspective is a surgical procedure. If an embryo is aborted in the seventh week, a living, organized group of cells that is not a recognizable form of the species has been removed from the uterus, and it ceases to exist as a living entity outside the uterus. Using words like *embryo* and *fetus* keeps things in the domain of medical procedures, where the issues are medical. Using the term *baby* imposes a different form of conceptualization. A *baby* is an independently existing human being, not just an unrecognizable group of the mother's cells that might be subject to a medical procedure.

Defenders of the morality of abortion are usually defenders of early abortion: abortion of the embryo or of a fetus well before it is viable outside the womb, usually in the first trimester. That is, they are defenders of the morality of removing from a mother a group of cells that is not an independent, viable, and recognizable human being.

Opponents of abortion use the word *baby* to refer to the cluster of cells, the embryo, and the fetus alike. The very choice of the word *baby* imposes the idea of an independently existing human being. Whereas *cluster of cells, embryo,* and *fetus* keep discussion in the medical domain, *baby* moves the discussion to the moral domain.

The issue of the morality of abortion is settled once the words are chosen. The purposeful removal of a cell group from the mother that does not constitute an independently existing, viable, or even a recognizable human being cannot be "murder." The word "murder" is not defined as such a medical procedure. The purposeful killing of a "baby"—an independently existing human being—can be "murder."

Is an abortion in the first trimester, say, in the seventh week, merely a surgical procedure that is morally neutral, and perhaps even moral if it is beneficial to the mother? Or is it the murder of a baby? The answer one gives depends on how one frames the situation and, correspondingly, on what word he or she uses—"embryo" or "baby."

When there is more than one possible framing of the situation, is there a right framing? Is there a single right way to conceptualize abortion? Both sides agree that there is a right framing, but they disagree as to which the right framing is. But both sides agree the answer is a moral one, and that it is a moral stance that chooses the right framing. These views are held with very deep moral conviction and sincerity on both sides; in many cases they are so deeply held that they are part of one's identity.

Interestingly, the choice of framing is not independent of politics and political morality. Catholics aside (religion is another, very relevant matter), liberals think of abortions and talk about them in terms of medical procedures, while conservatives think and talk about them in terms of killing babies. It could be that this is a historical accident, but I don't think so. I think it has everything to do with the two moral systems, those of the Strict Father and the Nurturant Parent.

What inclines me to this view is the attitude of pro-life advocates on the death penalty and on programs for reducing America's astronomically high rate of infant mortality, the highest in the industrialized world. Most pro-life advocates favor the death penalty. Most pro-life advocates do not support government programs for reducing infant mortality through, say, prenatal and postnatal care programs for impoverished mothers.

Indeed, some cynical liberals have even questioned the sincerity of pro-life advocates as not really being in favor of "life" as an absolute, since they support the death penalty. Other liberals have questioned the morality of pro-life advocates who want to save the lives of some unborn babies (those who would be aborted) and not save the lives of other unborn babies (the great many who die of inadequate pre- and postnatal care). To a liberal, it is both illogical and immoral for someone to want to save the life of an unborn baby whose mother does not want it, but not to want to save the life of a baby whose mother does want it.

I do not question the sincerity of pro-life advocates on the issue of abortion. I have met many of them, and their moral position seems completely sincere. But when I mention the death penalty, that is seen to be irrelevant and another matter completely. And when I mention pre- and postnatal care, they say they hadn't thought of it, but maybe it might be a good idea, though maybe not if the government does it. But they do not, with the fervor with which they oppose abortion, then go out and support pre- and postnatal care programs.

I do not think such opinions are either irrational or insincere. I think they are natural concomitants of having a conservative worldview, a Strict Father morality. Think back for a moment to the beginning of Chapter 9, where I laid out the major categories of moral actions for conservatives. So far as I can tell, those are the primary categories in terms of which conservatives naturally categorize actions as moral or not. Those categories do not mention life and death as primary criteria in themselves. Issues of life and death are seen as moral or not based on other criteria. The death penalty falls under Category 3, upholding the Morality of Reward and Punishment.

On the basis of those categories, conservatives support the death penalty and oppose social programs, for the reasons I gave above. Programs for pre- and postnatal care are social programs, and so they are opposed by conservatives. Moreover, conservatives would naturally assume that such care was the responsibility of parents. If poor parents could not afford adequate pre- and postnatal care, then it would be irresponsible for them to have children. For conservatives, the issue of infant mortality due to inadequate care is one of individual responsibility, not government action. It comes under moral action Category 2—self-discipline and self-reliance.

The death penalty and pre- and postnatal care simply fall, for these reasons, under different primary categories of moral action in the conservative moral system. But what about abortion? Why should conservatives tend to categorize as a "baby" what liberals see as an embryo?

Consider for a moment who is most likely to want an abortion. There are two classical kinds of cases: Unmarried teenage girls who have been having sex but have been careless or ignorant in the matter of birth control; women who want careers or independent lives and whose deepest aspirations would be destroyed by having a child at this point in their lives. There are, of course, many other kinds of cases—victims of rape or incest, for example, and women with a family already and neither the money nor the strength to raise more children. But the first two are the stereotypes.

Let us start with the first case. According to Strict Father morality, an unmarried teenage girl should not be having sex at all. It is a moral weakness, a lack of self-discipline, a form of immoral behavior, and she deserves punishment. She has to be responsible for the consequences of her actions if she is to learn from her mistakes. An abortion would simply sanction her immoral behavior. She would be "getting away with it." That is unconscionable, immoral—a violation of moral-action Category 2 (self-discipline and responsibility for one's actions).

Here is how Marvin Olasky, a major conservative writer, puts it (*Wall Street Journal*, March 22, 1995): "Unmarried lust and abortion go together like a horse and carriage. . . . men and women who shack up are nine times more likely to engender abortion than their married counterparts. . . . anything that increases promiscuity and discourages marriage . . . increases abortion."

Now the second case. In the Strict Father model of the

family, a woman's role is raising children. The moral order places men in leadership roles, not women. Women can work to help out the family and to help out men in business. But women should not be choosing careers or an independent careerist lifestyle over their natural role as mothers in a family. When a women chooses an abortion in order to place a career above motherhood, she is violating the moral order and challenging the entire Strict Father model. Abortion in such a case is immoral by moral-action Category 1 (self-defense of the Strict Father family model, where the father has authority) and Category 5 (upholding the moral order, with men ranked above women).

In both of the classical stereotypical cases, abortion violates Strict Father morality. Strict Father morality therefore incorporates very strong reasons for categorizing abortion as immoral. But categorizing abortion as immoral fits the "baby" frame, not the morally neutral medical frame. Thus, classical Strict Father morality correlates naturally with conceptualizing the object of abortion as a "baby"; once that is done, it becomes difficult if not impossible to classify abortion as anything but baby killing.

Once Strict Father morality chooses opposition to abortion and, with it, the use of the "baby" frame, that choice functions to reinforce the Strict Father model itself. A primary function of the Strict Father model is the protection of innocent children. Opposition to abortion provides an ideal opportunity to assert a protective function and justify Strict Father morality. What could be a more perfect model of a helpless, innocent child than a baby in the womb? What could be a more heinous crime than bloody murder? This reinforces Strict Father morality once you have it.

Once one classifies the object of an abortion as a "baby," the abortion becomes "baby killing" and it is natural for it to evoke deep and sincere moral outrage. Thus, as a conservative, it is completely natural to be morally outraged by abortion, to favor the death penalty, and to be opposed to government programs for pre- and postnatal care—all for different reasons.

One last thing. In this explanation, the crucial thing that forces an opposition to abortion and with it the use of the "baby" frame is the authority of men above women in Strict Father morality. In conservative feminism, as described below in Chapter 17, the authority of men over women is eliminated and the crucial condition for requiring opposition to abortion is not met. Thus, conservative feminists (both male and female) are not bound by the logic of their morality to be either pro-life or pro-choice. In short, the model predicts that there should be conservatives who are pro-choice, and that they should be those who do not rank men above women in the moral order.

At this point we must ask similar questions about liberals. Why do liberals support a woman's right to choose whether to have an abortion? Why is the liberal's object of nurturance the pregnant woman instead of the embryo or fetus or the cluster of cells? Why don't liberals use the "baby" classification, instead of the "fetus" one?

We saw that the conservative categories of moral action require conservative opposition to abortion in the two cases cited above. But the liberal categories of moral action work very differently. In Nurturant Parent morality, the teenage girl is "in trouble," she needs help and deserves empathy (moral-action Category 2—helping). The last thing she needs is somebody bawling her out, telling her how bad she is, and that she needs to be punished. Just being in this position is about as much punishment as a person should reasonably have to take. She isn't ready to be a mother, she has her whole life ahead of her, and, quite reasonably, shouldn't have her aspirations ruined (moral-action Category 4—self-development). She has plenty of time to have children and be a mother later when she can raise them properly. She should have the abortion if that's what she wants. There is nothing immoral about it.

Since there is nothing in the categories of liberal moral action that militates against the abortion and much that favors it, the categorization of the collection of cells as an embryo or fetus rather than as a baby is motivated, and, with it, the medical framing of the surgery. The story is pretty much the same for the career woman.

Liberals and conservatives, understandably, find each other's attitudes shocking. Conservatives, who cannot help but use the "baby" frame, find it hard to imagine anyone not using it and not thinking of abortion as baby killing. To a conservative, a woman making the choice to have an abortion is necessarily engaging in denial, finding an excuse for her own self-indulgence and for her immoral behavior and irresponsibility. Conservatives, given their moral system, cannot help but see it that way.

Liberals find the attitudes and actions of conservatives equally horrific. Conservatives are trying to ruin the lives of young girls in trouble and women trying to make a go of it in a man's world. They are hounding brave doctors and nurses who are working with diligence and courage to help those women. Their actions and their rhetoric encourage those who would threaten or even kill people who have dedicated their lives to helping women who desperately need them. They are encouraging a return to the days of dangerous back-alley abortionists. Given their moral system, liberals, too, cannot help but see it that way.

16

How Can You Love Your Country and Hate Your Government?

Versions of Strict Father morality and conservative politics are hardly confined to the United States. They are common throughout Europe and Asia. But American conservatism has a feature that seems peculiarly American and that often puzzles observers of American politics from other countries. It is the resentment toward government that often borders on, or extends to, hatred.

It is not that conservatives hate their country. Quite the opposite. They profess the highest order of patriotism; the superpatriots come from their ranks. Nor do they hate their system of government. Conservatives are clear on their allegiance to democracy. Nor do they hate the founders of their government. Their commitment to the founding fathers is unwavering. This is what puzzles foreign observers. How can conservatives love their country, love their system of government, love the founders of their government, but resent and often hate the government itself? This does not happen in most other countries with versions of conservative politics. You won't find it in France or Italy or Spain or Israel or Japan. Why?

Let us think about this question in terms of the model presented above. The Nation As Family metaphor turns this question about the nation into a question about the family: How can someone love their family, love the idea of the family, love their ancestors, but resent and hate their father? This, incidentally, does *not* imply that every conservative who resents or hates his government necessarily resents or hates his own father. But it does allow us to ask whether the general resentment of government correlates with something about the role of the father in the American Strict Father model in general.

Interestingly enough, there is a peculiar feature of the American Strict Father model, not found in most foreign models, to which such a resentment might be traced. It is that clause in the American Strict Father family model which says:

> The mature children of the Strict Father have to sink
> or swim by themselves. They are on their own and
> have to prove their responsibility and self-reliance.
> They have attained, through discipline, authority over
> themselves. They have to, and are competent to,
> make their own decisions. They have to protect them-
> selves and their families. They know what is good for
> them better than their parents, who are distant from
> them. Good parents do not meddle or interfere in
> their lives. Any parental meddling or interference is
> strongly resented.

Such a clause is not usually found in Strict Father models in most other countries—not in France, or Spain, or Italy, or Israel, or Japan. If we project this clause back onto the nation, using the Nation As Family metaphor, here's what it becomes:

> Mature citizens have to sink or swim by themselves.
> Citizens are on their own and have to prove their
> responsibility and self-reliance. They have become
> the leaders in their own family units (or local
> communities). They have to, and are competent to,
> make their own decisions. They have to protect them-
> selves and their families (or communities). They
> know what is good for them better than their govern-
> ment, which is distant from them. A good govern-
> ment does not meddle or interfere in their lives. Any
> governmental meddling or interfering is strongly re-
> sented.

Here we find many of the conservative attitudes toward the
federal government: it is distant, it doesn't know what's best
at the local level, it shouldn't meddle or interfere and is
resented if it does.

Here we have a remarkable form of explanation. A characteristically American part of the Strict Father family model
corresponds exactly to a characteristically American aspect
of conservative politics. What is explained is more than just
antipathy to government "meddling." It is the Principle of
Local Knowledge—that local governments always know better than the distant federal government how to spend their
money. This seems like just common sense. Of course it
isn't. There are plenty of cases where the reverse is true. A
hundred local governments acting separately cannot design
large water projects that benefit them all, or perform overall
ecosystem protection that requires coordinated action, or
plan and carry out a highway system. There are hundreds of
things better done at the federal level. What is interesting is
that the "common sense" persists and is so powerful. The
level of passionate resentment attached to this principle is
also interesting. It runs very deep.

The Abusive Father

It is not uncommon in this country for strict fathers (or mothers) to go too far and become abusive. Child abuse is a major problem in the country, as is child neglect. There is a syndrome of abusive distant fathers, often alcoholics, who are threats to their families, even after the children move away. Children often do have to protect their families from their fathers, as was the case in President Clinton's own family. Neglected children often see their parents as having no legitimate claims to authority over them.

You, yourself, may not have had an abusive or neglectful strict father or mother, but if you grew up in a community where parental abuse, neglect, and alcoholism were problems, you may well have acquired such a negative variation of the Strict Father family model in your conceptual system. There are at least two possibilities: (1) Your central Strict Father model may contain an abusive father. (2) Your central Strict Father model may be the ideal model discussed above, but the abusive Strict Father model may be a negative stereotype to be avoided.

The Nation As Family metaphor projects these two versions of an abusive Strict Father model onto two forms of possible conservative political views. In the first case, the government, like the abusive father, may be seen as inherently abusive, neglectful, ignorant, dangerous, and potentially out of control. This means that citizens have to protect themselves from their government, that the government may at any time come after them. It produces an antigovernment paranoia.

In the second case, there is the ideal of a government along ideal Strict Father lines: The government sets moral standards and keeps them, is responsible and protective, keeps its distance, doesn't interfere in your life, doesn't require money from you, and thus earns your respect. On the other hand, there is always the spectre of the abusive government to be guarded against, just in case.

Both of these views occur in the conservative community, and I have known people of both types. The first appears to be common among people who are survivalists or belong to the militia movement, at least those whom I have heard on interviews. Such people love their country as they would their family. They believe in their form of government, just as they believe in the family as an institution. They respect the nation's forefathers, just as they respect their own ancestors. But they resent and often hate and fear their present government.

Is this because they come from families or communities which had problems with abusive, neglectful, alcoholic fathers? Do they have a model—either a central model or a negative variation—of a family with an abusive or neglectful strict father? Are such models learned from generation to generation? Are they learned or are they propagated in any of our institutions: in the military? in competitive sports? in schools? in fraternities or other social organizations?

I do not know the answer to any of these questions. But they are questions that have to be asked, and I would not be surprised if the answer to some of them were yes.

Bigotry and the Moral Order

Exactly what is and is not in the Moral Order can vary from person to person. For example, the clause "Men above women" may or may not be present in the Moral Order hierarchy. If it is, it has the effect of sexism. Many of the clauses in the Moral Order correspond to forms of bigotry:

The racist clause: Since the dominant culture has been white, whites rank above nonwhites.

The anti-Semitic clause: Since the dominant culture is Christian, Christians rank above Jews.

The jingoist clause: Since this is an American culture where people born here have more power and status than immigrants, those born American rank above immigrants.

The homophobe clause: Since heterosexuality is dominant in our culture and homosexuals are stereotyped as weak, heterosexuals rank above homosexuals.

The superpatriot clause: Since America is the dominant country (the only superpower), America ranks above other countries.

If your conceptual system contains the Moral Order metaphor, as it will if you accept Strict Father morality, then your conceptual system contains the framework into which such clauses could fit; and, indeed, historically those clauses were present in the conceptual frameworks of many Americans. There was a time when they were all as American as apple pie. Many Americans have since dropped them, though they are still very much present for many others.

It is important to bear in mind that these define a "moral order." Those higher in the moral order are "better" and have a moral authority over those lower in the hierarchy. So, for instance, if all these clauses are in your hierarchy and if you happen to be a heterosexual white Christian American man, you are "better" than most people in the world.

According to the Moral Order metaphor, a just situation obtains when the Moral Order hierarchy is met in the world, that is, when men do have moral authority and power over women, when parents do have moral authority and power over their children, when human beings do have moral authority and power over nature, and so on. The bigoted clauses include whites having moral authority and power over nonwhites, and so on. In short, the Moral Order is the conceptual mechanism by which assumptions of superiority—and the moral standing of that superiority—are expressed.

We can now suggest an answer to the questions asked above. If conservatism is based on Strict Father morality, conservatives have the general metaphors of the Moral Order in their conceptual systems, with at least a couple of clauses, namely, God above human beings; human beings above animals and the natural world; adults above children; men above women. The Moral Order hierarchy used to have all the bigoted clauses in it; now it has many open slots available for such clauses.

Let us take an extreme example: the Ku Klux Klan. The Ku Klux Klan was bigoted in all the above ways. It professed the absolute superiority of white Christian American men. A way of expressing this in terms of the model would be that the KKK version of the Moral Order included all of the above clauses. It is important to recall that the Ku Klux Klan viewed itself as a moral organization, as religious and patriotic. What, in the eyes of its members, made it moral was that it upheld the Moral Order. They wore white to symbolize not only white supremacy but also morality, according to the common metaphor that Morality Is Light and Immorality Is Darkness. Their burning crosses symbolized morality via that metaphor and the superiority of Christianity in their moral order. They called themselves "knights" in a "royal order," which harked back to the superiority of the nobility over the commoners in the moral order. Back in the days when they rode horses, their riding of a horse symbolized the superiority of man over nature. Today, the superiority of man over nature in the Moral Order is symbolized by an ability to survive in the wilderness—hence, "survivalism."

The logic of the moral order included a notion of morality and justice; a situation is moral and just when the Moral

Order is met and immoral and unjust when it is reversed. The KKK was a vigilante group, one that saw it as a moral act to bring wrongdoers to justice itself. The KKK saw its job as righting wrongs of a particular kind—reversals of the moral order. If an uppity black or Jew or immigrant got too wealthy or powerful or haughty, the Moral Order was being reversed and it was the job of the KKK to set it right. The Moral Order, as they saw it, gave them the moral authority to hold their own courts and mete out their own punishments. The KKK operated according to a bigoted, vigilante version of Strict Father morality. It is no accident that they were political conservatives.

I have been speaking of the KKK in the past tense. But, in recent years, reports indicate that former KKK members have been joining the militia movement, which is well armed and is sanctioned by political conservatives. Why should such bigots find a natural home with militias?

The short answer is this: The militias see their job as upholding the Moral Order. The Moral Order is a hierarchy of legitimate authority. Any illegitimate use of governmental authority is thus seen as a reversal of the Moral Order. As we have seen above, those in the militia movement, like other conservatives, see liberal government policies such as the progressive income tax, gun control, environmental protection, and so on as interfering and meddling, as the illegitimate intrusion of authority into their lives; in short, as tyranny. The militias are preparing for vigilante action, if needed, to uphold the Moral Order against what they see as the illegitimate use of government power.

Conservative vigilantes, including those in the militia movement, are three steps from central conservatives. Here are the steps.

> Suppose (1) that the Strict Father in their model is too strict, that he is abusive. In the Nation As Family metaphor, that's an abusive, out-of-control, dangerous

federal government—a Big Daddy you have to defend yourself against.

Suppose (2) that instead of just a normal conservative's resentment at the federal government, one has what one sees as justifiable rage against the government's illegitimate use of power.

Suppose (3) that one believes in vigilantism, namely, that one believes that it is moral to be an agent of justice when no one else will do the job.

Let the rest of such a model just be that of central conservatism.

These three differences make vigilante action seem moral. They could justify acts like the shooting of abortion doctors. These differences might seem like enormous differences from central conservatism, but they are differences of degree—and not necessarily large differences of degree.

First, abuse is strictness that has gotten too strict. But how do you draw a line between strict and too strict?

Second, continuing deep resentment at the federal government is not that far from rage against the government; again they are on the same continuum and not all that far apart.

Third, sanctioning vigilantism of a violent sort is on the same continuum as sanctioning violent protests at abortion clinics. Both are performed with a sense of self-righteousness and moral indignation, and are seen as forms of moral action.

In short, the models are basically the same and differences are differences of degree. They are, of course, extremely important differences of degree. Nonetheless, there is a slippery slope from one model to the other, from normal law-abiding conservatism to violent conservative vigilantism.

That slippery slope raises a set of questions:

Is the conservative vigilante model just a more extreme variant of the normal conservative model?

Does perpetuating the normal conservative model also perpetuate the conservative vigilante model?

Will stirring up moral outrage at liberals tend to move people from the normal conservative model to the conservative vigilante model?

Both liberals and conservatives have clear answers to these questions—opposite answers. Conservatives say no, liberals say yes. To liberals it is just common sense. Every conservative that I know or have heard discussing these issues in the media has disagreed. Part of this may just to be a matter of defending one's own territory. But there are other reasons for this disagreement.

Liberals, who look for social causes of immoral behavior, have argued that it is often the case that conservative morality and conservative politics themselves can be social causes of immoral behavior. Another social cause can be the whipping up of moral outrage against liberals and the government by conservative talk-show hosts proselytizing for the conservative moral position.

But from a conservative worldview, this can only be nonsense. To the conservative, immoral behavior is attributable to individual character, not to social causes: What is right and what is wrong are clear, and the question is whether you are morally strong enough to do what's right. It's a matter of character. Conservatives believe that if an extreme conservative commits a crime, say killing people, in the name of vigilante justice, then conservatism itself cannot be held to blame, nor can those who spew hate over the airwaves. The explanation instead is that that individual had a bad character, that is, a bad moral essence; either that or he was crazy, craziness being a different kind of moral essence. "He was a bad person" is not only a sufficient answer; it appears to be a necessary answer, since explanations on the basis of social causes are excluded.

Part Five

Summing Up

17

Varieties of Liberals and Conservatives

Liberalism and conservatism are anything but monolithic. Both provide rich moral and political worldviews, rich enough to permit a wide range of variation—an almost dizzying complexity that if looked at from a distance might just seem like a big soup. But when we look closely, we see a great deal of systematic variation.

The kind of complex variation we find is just what anyone who studies human categorization would expect. Most categories are similarly complex. In my book on categorization, *Women, Fire, and Dangerous Things* (A2, Lakoff 1987), I surveyed the evidence showing that one of the main sources of such complexity is "radial" category structure. A "radial category" has a central model, which gives rise to systematic variations that radiate out from the center like the spokes of a wheel. The example I gave in Chapter 1 of this book was the category "mother," where the central category was defined in terms of a cluster of models—birth, genetic inheritance, nurturance, and marriage. This gives rise to many variations within the category of mothers: birth mothers, single mothers, unwed mothers, working mothers, stepmothers, foster mothers, genetic mothers, surrogate mothers—all variants based on the classic central case: mom, who gave birth to you, gave you half your genes, raised you, and is married to dad. This doesn't make the classic mom a "better" mother, or "more of a mother." All it means is that good old mom is a prototype, a central case, and that the variants are defined with reference to her.

Most of our categories are complex in this way, containing a great many variations on a central case. The categories "conservative" and "liberal," one would expect, should be no different. And it appears that they do show the same kind of radial structure that other categories show.

Throughout most of this book I have been concerned to describe the central models of liberals and conservatives. But since noncentral cases vary from the central cases, most readers who are either liberal or conservative should have found places in the discussion above that did not apply to them and to many of their friends. I discussed that in advance in Chapter 9 and gave a promissory note that we would look at the parameters of variations later. The time has come to pay off that promissory note.

We saw in Chapters 5 and 6 that the Strict Father and Nurturant Parent models were central members of radial categories, and that there were four parameters for the variations on the central members: (1) Pragmatic-Idealistic; (2) Linear Scales; (3) Moral Focus; and (4) Clause Variation in the Moral Order metaphor. I have argued over the past seven chapters that the Nation As Family metaphor, when applied to the central model in the Strict Father and Nurturant Parent moral systems, yields central forms of conservatism and liberalism. I will now argue that the same parameters of variation applied to the central models systematically characterize varieties of conservatives and liberals.

Before we move on to the details, it is important to consider the role of the study of variations within the overall theory presented in this book. If we can make sense of a variety of types of liberals and conservatives—of the views that they hold and the reasoning they use—then this analysis of variations lends greater support to the analysis of the central models. The reason is this: Variation data are extraordinarily complex. The theory of radial categories claims that variations should be systematic in a certain way, determined by the application of parameters of variation to the central model. But the account of variations only makes sense given the central models. Thus, support for an account of existing variations is also support for the existence of the central models on which the variations are based.

Let us now turn to the details.

Parameters of Variation

Let us begin with the parameters of variation described in Chapters 5 and 6.

1. Linear Scales
2. The Pragmatic-Idealistic Dimension
3. Moral Focus
4. "Clauses" within the Moral Order (Strict Father model only)

Since the political models arise from the family-based models through the Nation As Family metaphor, we should expect to find the same parameters of variations in the political models—and we do. Let us begin with the Pragmatic-Idealistic dimension.

Pragmatic Variations

As we saw in Chapters 5 and 6, pragmatic variations on the idealistic central models arise in the following way. In both central models, the pursuit of self-interest serves as a means to an idealistic end: the end of self-reliance in the Strict Father model and the end of nurturance in the Nurturant Parent model. The pragmatic variations on these models reverse means and ends, making the pursuit of self-interest an end in itself and making either nurturance or self-discipline and self-reliance into two very different means toward that end. Through the Nation As Family metaphor, the pragmatic variations on the family models are projected onto politics. What results are pragmatic versions of conservatism and liberalism.

Pragmatic conservatism is less idealistic than central conservatism. The goal is to get ahead, to serve your self-interest. The idealistic parts of conservatism are seen as effective means to achieve that goal. If you want to get ahead, you'd better be self-disciplined and self-reliant. You'd better stick to the straight and narrow so that nobody gets upset with you. You'd better obey legitimate authority, or you could get hurt.

A homelier way of seeing the difference is to ask which is the dog and which is the tail. In central conservatism, the idealistic dog wags the self-interested tail. In pragmatic conservatism, the self-interested dog wags the idealistic tail.

Pragmatic conservatives are more likely to compromise on the principles of Strict Father morality for the sake of self-interest. Central, or idealistic, conservatives are more likely to stick to their principles, even when it goes against their self-interest. Of course, since self-reliance is one of those principles, they can't go against self-interest too much.

As should be obvious, the pragmatic-idealistic distinction is not absolute but a matter of degree, since it can vary from time to time and from issue to issue. You can also choose issues that are beyond compromise, issues where it would take a lot to compromise, and issues where you are willing to compromise. A great many variations on conservatism result from a decision as to where to be idealistic and where to be pragmatic.

Liberalism has a similar idealistic-pragmatic dimension: In central (or idealistic) liberalism, the pursuit of self-interest (say, for money and power) is a means toward the end of nurturant morality: to be better able to help others and promote fairness and nurturance in society.

In pragmatic liberalism, the end is to allow people to best pursue their self-interest, and the means is nurturance: empathy, being in a nurturant environment, being self-nurturant, being basically happy, being treated fairly and treating others fairly. There are two cases here.

First, someone can best seek self-interest as an end if he is an object of nurturance, if people empathize with him, help him when he needs help, and help him fulfill his potential, if he is allowed to be basically happy, and if he is treated fairly.

Second, someone can best seek his own self-interest if he empathizes with others, helps others, takes care of himself, is basically happy, and treats others fairly. The idea is that being both an object and agent of nurturance helps you pursue your self-interest as an end in itself.

Thus, pragmatic liberals see social programs as a way to help others pursue their self-interest, while idealistic liberals see social programs as a commitment to providing basic human needs, which is an end in itself. To pragmatic liberals social programs are investments; to idealistic liberals, they are a matter of civic duty.

Again, it is a case of whether the self-interested dog wags the idealistic tail (pragmatic liberal), or whether the idealistic dog wags the pragmatic tail (idealistic liberal).

As in the case of pragmatic conservatives, pragmatic liberals are more willing than idealistic liberals to compromise on their principles if it serves either their self-interest or the self-interest of others. And as in the case of conservatives, liberals can't compromise too much or people will stop helping them, which would hurt their self-interest. Thus, in both cases, there is a self-regulating mechanism that keeps pragmatic liberals and conservatives from losing touch with their respective moral systems.

Pragmatic political figures are sometimes called "moderate" or "middle of the road" because of their willingness to compromise. But the term "moderate" gives the false impression that there is a linear political continuum, with people distributed along it. The continuum metaphor hides the major role played by moral systems and the fact that pragmatic politicians in America are usually pragmatic versions of either liberals or conservatives.

Not surprisingly, both pragmatic liberals and conservatives are commonly criticized by their more idealistic colleagues for betraying the moral principles of their respective ideologies, for "waffling" on issues, or for watching which way the wind is blowing.

LINEAR SCALES

In Chapter 5, we saw that the Abusive Parent model is an extreme version of the Strict Father model, in which punishment is very severe. The difference is one of linear scale, and the difference of degree results in a difference of type. We saw the same in the previous chapter in the analysis of conservative vigilantism, which differs from central conservatism by taking extreme positions along three linear scales: (1) the degree of punishment in the Strict Father family model; (2) the degree of resentment and anger at the meddling parent in the Strict Father family model; and (3) the degree to which violence is sanctioned.

In Chapter 6, we saw that virtually every aspect of the Nurturant Parent model is subject to linear scale variation. Many of those family model variations correspond to variations on liberalism. Take, for example, an overprotective parent who puts too much energy into protecting children when no protection is needed. Conservatives see "excessive" government regulation as a form of overprotectiveness. Or take a parent who puts excessive energy and other resources into nurturance, so much so that he doesn't take proper care of himself. This corresponds to the conservative criticism that liberals spend so much money taking care of people that they will bankrupt the treasury. In this way, family model excesses are projected onto what are seen by conservatives as political excesses.

Moral Focus

Thus far we have seen two kinds of variations within the liberal and conservative categories: (1) pragmatic-idealist variations, and (2) linear scale variations. A third kind of variation is one of moral focus. Let us begin with liberals. It is hardly uncommon for blacks, whether ordinary citizens or politicians, to focus on race, or for women to focus on gender issues, or for members of ethnic minorities to focus on ethnic issues, or for gays to focus on gay rights issues. These kinds of focus constitute a politics of identity. Other kinds of moral focus are issues like the environment, or poverty, or labor relations, or education, or health care.

What exactly is moral focus? It is giving moral priority to one particular domain of interest over others. The result is that one domain is seen as having primary moral significance over other domains.

Thus, people can be liberals with the same family model, the same moral system, and the same degree of idealism vs. pragmatism, and yet, as political beings, they may live in different universes. If the most important thing to you is race, you may have little interest in environmental issues, except perhaps as they apply to race. The views of a liberal on issues may vary greatly depending on his focus. For example, consider two liberals, one with a moral focus on civil liberties and the other with a moral focus on violence against women. The first will probably support keeping pornography legal, while the second may want to ban it. Or consider two liberals, one with a moral focus on race and the other with a moral focus on environmentalism. The liberal focused on race may be against any environmental regulations that would cost blacks jobs. Black people alive now may be more important to him than owls or old-growth forests.

Moral focus is very different from self-interest, with which it is often confused. A white person may have the rights of nonwhites as a moral focus, as many whites did during the civil rights movement. They didn't do it out of self-interest. Some blacks at the time may have been for civil rights out of self-interest, but I suspect that, for most blacks, the issue of race was primarily a matter of moral focus: the lens through which they viewed the world and the domain in which their moral system was of most immediate relevance for them. Environmentalists rarely if ever have the environment as a moral focus merely out of self-interest. Instead, they find the environment to be the locus of major moral relevance in their lives.

Individual vs. Social Focus

In the Nurturant Parent model of the family, children have the right to have their basic needs met and the right to fair treatment by their parents. Children also acquire family responsibilities as they grow up. The Nation As Family metaphor translates these aspects of family life into aspects of political life: The individual rights and social responsibilities of citizens. Either can receive a moral focus.

Not surprisingly, there is a version of liberalism with a moral focus on individual rights in the domains of politics and economics. It sees government as providing and protecting such rights. Let us call it *rights-based liberalism*. (See References, C3; Rawls 1971). Correspondingly, there is a form of liberalism with an opposed moral focus, a focus on social responsibility. Let us call it *communitarian liberalism*. (See References, C4; Etzioni 1988, 1995). Each of these, of course, has many versions.

MORAL DEFOCUSING: INTERNAL AND EXTERNAL ORIENTATION

Just as one can focus on one aspect of a model, giving it highest priority, so one can defocus on an aspect of a model, taking priority away. The differences between focusing and defocusing on an aspect of a model can be striking.

Consider, for example, what I will call the "internal" aspects of Nurturant Parent morality: self-nurturance, self-development, and moral happiness. There are liberals who have made one or more of these internal aspects their moral focus.

The human potential movement, for example, is concerned with just these aspects of life, which are seen at least as a moral concern, and often as a political concern. This includes the exploration of alternative forms of religion—various forms of Judeo-Christian renewal in the direction of Nurturant Parent rather than Strict Father biblical interpretations, as well as exploration of eastern religions and forms of nature worship. The politics of religion is very much an issue in many of these groups. They realize that spiritual life is intimately tied to moral life, which is intimately tied to political life.

A moral focus on self-nurturance and moral happiness has produced a moral and political movement centered on healthy and good-tasting food: organic produce, organic farmers' collectives, and vineyards are supported by networks of chefs and restaurants who see themselves as involved in a politics of nurturance. From this perspective, cooking healthy, good-tasting, and preferably organic food is part of a politics that overlaps with environmentalism. The development of farmers' markets is part of this; fresh produce is best grown locally and brought to market every day or two. If you want fresh fish, then local streams and coastal waters will have to be kept clean and communities of fishermen will have to be protected. All of these are political issues. And they are set within an entrepreneurial capitalist framework—developing tastes and markets and means of making such enterprises profitable as well as socially useful.

A moral focus on all these "internal" aspects of nurturance leads to a morality and politics of art and education. Bringing art, crafts, design, and all sorts of aesthetic considerations into everyday life is a contribution to self-nurturance, self-development, and moral happiness, and is thus a contribution to the idea of a nurturant society. The same is true of an education in which art, health, and the study of many cultural and spiritual traditions have central roles.

These "internal" aspects of Nurturant Parent morality can be defocused as well, and there are a great many liberals for whom the internal aspects of nurturance are absent from both their morality and their politics. Such liberals are "externally oriented," oriented toward social issues alone. Externally oriented liberals tend to be more ascetic than internally oriented liberals. They tend to focus more on economics, race, class, and gender. Some have no internal focus at all and have no understanding of internally oriented liberals. Indeed, they may misinterpret internally oriented liberals in the same way that conservatives do, as narcissistic, self-indulgent, and hedonistic.

Many liberals have a balance of internal and external aspects, with no particular focus on either. But there are some who are either entirely internally oriented or externally oriented, and they may have a hard time talking to one another.

MORAL FOCUS IN CONSERVATISM

Conservatives too have various kinds of moral focus. James Dobson's Focus on the Family group, not surprisingly, has the family as its moral focus. The pro-life movement has abortion as its moral focus. Other common forms of conservative moral focus are violent crime, opposition to affirmative action, moral education, illegal immigration, welfare, the right to bear arms, and so on. These give rise to as many varieties of conservatism as there are varieties of liberalism.

Libertarians

Libertarians provide a very interesting challenge to the study of variations on a central model. Libertarians see themselves as forming a separate political category, neither liberal nor conservative, but something unto itself. An analysis in terms of variations on central models suggests that their view of themselves is not entirely accurate.

Suppose we start by looking at the central conservative model. Consider a variant on that model that is pragmatic in the extreme, that is, think of a conservative who sees the pursuit of self-interest as the principal end, and conservative morality (self-discipline, self-reliance, etc.) as a means to that end. Someone who is extremely pragmatic will be willing to sacrifice aspects of conservative morality if it interferes with the pursuit of self-interest. Now imagine such a pragmatic conservative having the moral focus: noninterference by the government. So far as I can tell, this is what a "libertarian" is, namely, an extremely pragmatic conservative whose moral focus is on noninterference by the government. In short, a libertarian is two steps away from a mainline conservative.

Such a person will believe that free enterprise should be as unrestricted as possible and that people should be self-disciplined and self-reliant in order to pursue their self-interest. He will be very much against social programs, taxation, government support of education and the arts, government regulation, and gun control. But the libertarian's moral focus on noninterference by the government and his extreme support of the pursuit of self-interest will make him a radical advocate of civil liberties. He will oppose any governmental restrictions on free speech, pornography, abortion, homosexuality, and so on. He will probably support the rights of women, gays, and minorities to equal opportunity, but be strongly against affirmative action on the grounds that it gives individuals things they haven't individually earned. He will most likely be pro-choice on abortion, but not believe that the government should pay for abortions. And since he gives priority to the pursuit of self-interest over the rest of the conservative moral system, he will not have the moralism of mainline conservatives; the seven deadly sins may not be sins for him.

A good example would be drug addiction, which, to many libertarians, would not in itself be immoral. Libertarians commonly favor the decriminalization of drug use and sale on the grounds of maximum noninterference by the government and maximum pursuit of self-interest. They frequently argue that government interference in the drug trade has artificially driven up the price of drugs, brought criminals into the drug market, and forced drug addicts to turn to crime to support their habits. Decriminalization, they argue, would allow honest business to pursue the drug trade, bring in competition, lower prices enormously, not force users to turn to crime, and not make it profitable enough for major crime syndicates to bother with.

The libertarian's advocacy of civil liberties will bring him into overlap with liberals on many positions. But the source of that advocacy comes from a different place—from a conservative model with minimally restricted pursuit of self-interest and a moral focus on noninterference. The advocacy of civil liberties in Nurturant Parent morality comes from the nurturance model, especially the concern with empathy, with fair distribution, with happiness, with development of one's potential, and so on. Empathy and fair distribution are not libertarian concerns.

The fact that libertarians and political liberals both strongly advocate civil liberties is a superficial similarity. They do so for very different reasons, out of different moral impulses, with a very different spirit. Though two steps away from mainline conservatism, libertarians are conservatives in three very important respects: (1) Their concern with noninterference by the government comes directly out of conservatism, out of the idea that the government is inappropriately paternalistic, that mature citizens should be left to take care of themselves. (2) They preserve primary conservative moral priorities: self-discipline, self-reliance, and individualism, rather than the cultivated interdependence required by the nurturance model. (3) They do not give priority to the values of Nurturant Parent morality: empathy, nurturance, interdependence, fairness, and responsibility for others.

There are, of course, lots of variations possible within the category of libertarians. One would no more expect uniformity there than in any other radial category. But variation within the ranks of libertarians is not random. One source of variation is the degree to which a given libertarian preserves conservative moral positions; for example, some libertarians might echo the conservatives' aversion to drugs because drugs arise from, and perpetuate, moral weakness. In general, the variation among types of libertarians reflects their conceptual links with conservatism. We don't tend to find libertarians supporting welfare or the progressive income tax or government protections of various kinds.

Thus, despite the claims of libertarians to be a category unto themselves, they appear to be just two steps—two important steps—from central conservatism, and the variation within their ranks seems to tend toward conservatism. There is, after all, a reason why the scholars at the libertarian Cato Institute seem largely to be writing in support of conservative rather than liberal positions. Nonetheless, there is no objective answer here. They are far enough away to think of themselves as a separate category and close enough for others to think of them as conservatives.

Liberal Strict-Father Intellectuals

I remarked in Chapter 1 that we do not all, or even mostly, have a coherent politics, that one might, for example, be conservative on foreign politics and liberal on domestic politics. The descriptions of Strict Father and Nurturant Parent moralities characterize just what it means to be liberal or conservative on an issue or an area of policy: it means to apply a given family-based moral model to a domain of politics through the Nation As Family metaphor.

But it is also possible for someone to be a strict political liberal, applying only the nurturance model in his politics, while being a conservative in other aspects of life. A familiar case is a class of liberal intellectuals who apply the Strict Father model to their intellectual lives.

Consider someone who is a thorough going liberal, but whose intellectual views are as follows:

> There are intellectual authorities who maintain strict standards for the conduct of scholarly research and for reporting on such research.
>
> It is unscholarly for someone to violate those standards.
>
> Young scholars require a rigorous training to learn to meet those scholarly standards.
>
> The only way they can learn appropriate scholarly rigor is to be given difficult assignments and held to

a high standard of performance, for example, to be given difficult tests and graded harshly.

Students require the incentives of grades if they are to develop the self-discipline needed to be a scholar. High grades are rewards and low grades are punishments. Receiving consistently high grades is a sign of self-discipline and therefore of good scholarship.

Students should not be "coddled." They should be held to strict scholarly standards at all times.

The goal of scholarly training is to produce rigorous scholars who are self-disciplined and self-reliant, that is, who can maintain scholarly rigor and uphold scholarly standards on their own.

This is an application of Strict Father morality to academic life. Here academic scholarship is conceptualized metaphorically as a version of Strict Father morality. The conceptual metaphor can be stated as follows:

Academic Scholarship Is Strict Father Morality.

- Mature Scholars Are Strict Fathers.
- Intellectual Authority Is Moral Authority.
- Scholarliness Is Morality.
- Unscholarliness Is Immorality.
- Scholarly Rigor Is Moral Strength.
- Lack of Scholarly Rigor Is Moral Weakness.
- Scholarly Discipline Is Moral Discipline.
- Scholarly Standards Are Moral Standards.
- Students Are Children.
- Teaching Is Setting Rules for Moral Behavior.
- Good Grades Are Rewards for Moral Behavior.
- Bad Grades Are Punishments for Immoral Behavior.
- The Prospect of a Good Grade Is a Moral Incentive.
- Tests Are Tests of Moral Strength and Self-Discipline.

- Scholarly Success Is an Indicator of Good Moral Character.
- Scholarly Failure Is an Indicator of Bad Moral Character.

Among the entailments of this metaphor are:

> Intellectual authority must be followed; it is not only unscholarly to do otherwise, but it violates the system of academic authority as defined by Strict Father morality.
>
> Competition for grades builds character and is an incentive for good scholarly practice.
>
> Scholarly achievement is an individual matter, a measure of an individual's moral worth.
>
> Students who are intellectually weak should be allowed to fail; only if they are punished can they learn self-discipline.
>
> Coddling or indulging a student will make him intellectually weak.
>
> Grades are a measure of a student's intellectual worth.

Much of the academic world and academic institutions are run according to this metaphor, which is based on Strict Father morality. Intellectuals who accept this view of the academic world may be political liberals, but they are intimately acquainted with Strict Father morality and practice it in their everyday professional lives.

Feminisms

I would now like to turn from varieties of liberalism and conservatism to varieties of feminism, for a number of reasons. First, an adequate theory of the variations within categories must be able to account for the varieties of feminists.

Second, we must be able to account for the existence and nature of conservative feminists as part of that theory of variations.

But lastly, and perhaps most importantly, feminism is a major part of the liberal political scene, yet there is so much variation within feminism that it is often hard to make sense of all the variants. For this reason, it is important to show that the mechanisms for characterizing variation that we have been discussing can make sense of a highly complex area of politics.

GENDER

There is a big difference between sex (a biological concept) and gender (a cultural concept). Gender is characterized by a collection of common folk models about sex roles. Each such folk theory characterizes a single stereotypical property of men and women. Taken collectively, the folk models characterize stereotypes of what is masculine and feminine, so that the stereotypical male is masculine and the stereotypical female is feminine. Here is Alan Schwartz's (References, A2, Schwartz 1992) account of those folk models.

The Physical Prowess Model
    Men are strong.
    Women are weak.
The Interaction Model
    Men are dominators.
    Women are cooperators.
The Family Roles Model
    Men are providers.
    Women are nurturers.
The Division of Labor Model
    Men work in the public sphere.
    Women work in the domestic sphere.

The Thought Model
  Men are rational, objective, detached.
  Women are emotional, subjective, in touch with
    themselves.
The Sexual Initiation Model
  Men initiate sexual behavior.
  Women respond.
The Discourse Model
  Men talk in order to act on the world.
  Women talk to maintain social networks.
The Morality Model
  Men have a morality based on laws and strict rules.
  Women have a morality based on nurturance and
    social harmony (References, B2, Gilligan
    1982).

In each case, the property ascribed to men has a higher social value.

Masculine gender is defined by the collection of stereotypical male properties in these models; feminine gender, by the collection of stereotypical female properties. Gender is therefore characterized in terms of cultural roles, not biology. Hence, there can be feminine men and masculine women.

Most of the forms of feminism that have developed over the past thirty years have been set within a liberal context, and for a good reason. Liberalism allows for social causes, and gender stereotypes are social and seen as having causal powers. Since the male roles in gender stereotypes are more highly valued in society, gender stereotypes are seen as giving power to men. Feminism in a liberal context sees this as unfair and believes that such unfairness should be eliminated. Since liberalism is concerned with social causes and fairness, such views are fundamentally liberal.

Within liberalism, there is thus a general form of feminism. Feminism (1) assumes that the above-mentioned gender stereotypes exist; (2) assumes that higher values are placed on the male roles; (3) assumes that these values given to social stereotypes have causal powers that give men a dominant position in society; (4) sees male dominance as unfair and to be remedied.

Given this general form of feminism, specific forms arise as a result of (1) differences of moral focus, and (2) differences of opinion about the truth of the gender stereotypes. Here are some examples:

RIGHTS-BASED FEMINISM

There is a feminist version of rights-based liberalism that (1) takes individual rights as a moral focus and (2) denies the validity of the gender stereotypes. The result is *rights-based feminism*, which sees government as the appropriate remedy for unfairness to women in the political and economic spheres. On this view, it is the role of the government to eliminate political and economic unfairness to women. The Equal Rights Amendment seeks to commit the government to this task generally. The National Organization of Women, NOW, is an organization of rights-based feminists.

RADICAL FEMINISM

There is a form of liberalism called *radical politics* which has as its moral focus equality in power relations in every domain of life. The feminist version of this is called *radical feminism*, which (1) denies the validity of the gender stereotypes and (2) has a moral focus on strict equality of power relations between the sexes in all domains of life. It claims there is, or should be, no cultural difference between men and women that results in any power differential, that power differentials between men and women should be eliminated in every sphere of human activity, and that this can be done only by individuals, not the government.

BIOCULTURAL FEMINISM

There is a third form of feminism that accepts the truth of some or all of the gender stereotypes, but believes that feminine gender roles should have a social value as high or higher than masculine ones. In other words, society should place equally high or higher value on cooperation, nurturance, emotionality, the domestic sphere, the maintenance of social networks, nurturance, and social harmony. This view is called *biocultural feminism*. Biocultural feminism takes nurturance itself and the biological nature of women as moral foci. Biocultural feminists believe that nurturance is a feminine gender trait and that it is a higher, more moral basis for society than domination, which is seen as a masculine gender trait. They believe that women's social roles (those involving nurturance) need to be valued at least as highly as, if not more highly than, men's (which involve dominance).

Different forms of biocultural feminism arise from taking different moral foci. One major moral focus is on ecology; that results in eco-feminism. Another major moral focus is on spirituality. This sees the Judeo-Christian tradition as dominated primarily by men and masculine values. It sees the Judeo-Christian tradition as having failed to provide for women's spirituality and for a nurturance-based morality. One variety of the women's spirituality movement seeks to reform existing Judeo-Christian religions by giving them a focus on nurturance (see Chapter 14 above). Others seek to create new religious movements, for example, the goddess movement, in which the earth is seen as a nurturant mother and a divine being. Another variant on this theme is wicca, which focuses on the power of women's spirituality and women's special ability to use their spiritual power to serve the cause of nurturance.

When moral focus is placed on the sex act itself, the result is *lesbian feminism,* another form of biocultural feminism. In lesbian feminism, lesbian sex is seen as being centered on nurturance, while heterosexual sex is seen as centered on domination. Though lesbians prefer nurturance in sex and women as sex partners, lesbian feminism is not necessarily antimale (though it may be). Many lesbian feminists simply want to bring nurturance and nurturant gender roles into prominence in American culture in general and in sex in particular.

What we see in these variants of feminism within liberalism is the generalized feminism discussed above, and variants of it defined by two kinds of parameters: (1) acceptance or nonacceptance of the gender stereotypes and (2) moral focus.

CONSERVATIVE FEMINISMS

On the same page on which Rush Limbaugh uses the term "femi-Nazis," he says "When I attack feminism, I am not opposing equal opportunities for women. I am totally in favor of equal pay for equal work" (C1, Limbaugh 1993, p. 233).

On the whole, conservatives are opposed to feminism. Yet there is emerging a generation of conservative women who see themselves as feminists; at least, they believe that women are strong, should have equal opportunity, and deserve equal pay for equal work. If feminism were no more than that, Rush Limbaugh would be a card-carrying feminist. The question of what a conservative feminist is, then, is important first because there are going to be many more of them, and second, because it sheds light on what conservatism is, what feminism is, why feminists have tended to be liberals, and why conservatives are appalled by the classical liberal varieties of feminism.

There are certain things that conservative feminism cannot be. Conservatives do not believe that there are social causes of individual failures, as we have seen. They believe that, if you have enough self-discipline and character, you can succeed. Therefore, conservative feminists cannot accept the idea that gender stereotypes have social causal powers. Nor can they accept the idea that social programs such as affirmative action are needed to "remedy" a social cause that they do not believe, on principle, could possibly be responsible for individual failures. A conservative feminism must be a version of Strict Father morality.

What I have been calling Strict Father morality is what many feminists mean by the term "patriarchy." I have not used the term, partly because of its negative and ideological overtones and partly because I wanted to be more specific. Within Strict Father morality, the part that applies specifically to women is the metaphor that the Moral Order Is the Natural Order (of dominance): God over human beings; human beings over nature; parents over children; men over women. It is this metaphor that plays the role in Strict Father morality of justifying the moral authority of men over women, both in the family and in society at large.

Now, the Moral Order in the Strict Father worldview has changed in Western culture over the years. It used to include whites over nonwhites, for example, and nobles over commoners. That has changed very considerably. Suppose, now, that the change in the Moral Order metaphor moves one step further and eliminates the clause "men over women." What you get is a kind of conservative feminism. It keeps all aspects of conservatism except those tied to men's domination and moral authority over women.

Here's what changes: In the family, men and women have equal responsibility for decisions. Men no longer have a say about women's sexuality. Options are open. If a woman chooses to have sex before marriage, it is her own business, provided she is no longer dependent on her family and provided she takes full responsibility for the consequences of her actions. There is no pressure for women to put homemaking ahead of a career. And there is no reason, as a feminist, not to be sexy and sensual, not to use femininity to gain power, not to dress well, and so on. Conservative feminists feel comfortable about the use of power, including sexual power.

On most issues, a conservative feminist is the same as other conservatives. She is against affirmative action for the same reason all conservatives are: because it gives people something they don't earn. Unrestricted free enterprise supports, in principle, equality of opportunity for women and equal pay for equal work. But affirmative action to "level the playing field" would still be immoral. She would still believe in moral authority, in a reduced version of the moral order, in hierarchy, and in elites of achievement. But women would not be lower in the moral order just by virtue of being women. She would still believe that punishment and reward are the basis of morality, and she would still believe in the primacy of moral strength (self-discipline, responsibility, and self-reliance), and she would still be against welfare and other social programs, against gun control, for the death penalty, against government regulation, and so on. Because she believes in the primacy of moral strength, self-discipline, and responsibility, she would have little patience for women who see themselves as victims, say, of date rape. They know the score and, with self-discipline, can just say no. She would look down on women who whine; they are morally weak and give women a bad name. A conservative feminist would not be constrained by her moral worldview to be either for or against abortion. She could go either way on that issue.

I have encountered women with such views and I believe
there is no lack of them. In political life, Governor Christine
Todd Whitman of New Jersey is an example. A good example among conservative authors would be Lisa Schiffren, a
former speechwriter for Dan Quayle. Here, as an example
of a conservative feminist position, is part of Schiffren's
argument *against* making the men who impregnate unmarried women on welfare financially responsible for the care
of their offspring. Part of the argument is that such a policy
won't work. Another part is that "It is bad policy for the
state to enforce a contract that does not exist." But the main
argument is that women should be responsible for their own
actions.

> Since women and girls have sexual autonomy, they
> can and should be held responsible for how they use
> it. Before I am accused of blaming the victim, or
> wishing to deny women sexual freedom, recall that
> the women in question are not the classic victim cari-
> catures that the feminist-welfare lobby likes to cite.
> These are not wives bound by law or financial depen-
> dence to husbands. They are single women who con-
> trol economic resources, in this case the A.F.D.C.
> check.
>
> Contraception to prevent pregnancy is available—
> including Norplant, Depo-Provera and the pill. Abor-
> tion is an option.
>
> Girls have the same educational opportunities and
> most of the same economic opportunities boys have.
> This makes the choice of dependence less acceptable
> for poor women, just as it has for middle-class
> women.
>
> The most useful thing we can do for girls on the
> verge of becoming welfare mothers is to make educa-
> tion, work, and marriage preferable to subsisting on a

welfare check. (*New York Times,* Op-Ed, August 10, 1995)

Here is a classic conservative argument with a feminist twist: women with sexual freedom, educational and economic opportunity, controlling their own fates and therefore being responsible for themselves.

THE CONSERVATIVE GODDESS MOVEMENT

Is it possible for conservatives—women and men—to see the earth as a goddess? Could there be a conservative spiritual eco-feminism? In fact, such a variation on conservatism is already in existence—and it has arisen out of evangelical Protestantism. It accepts the truth of the gender stereotypes, as does the goddess movement. But it does not want women's values to replace men's in society. It accepts the Strict Father model of the family, the limitation of women to the domestic sphere, and the role of the wife as functioning to raise the children and support her husband's authority. In short, it supports the existing moral order and the legitimate authority of men as leaders. And yet, it is a version of the goddess movement, complete with solstice rituals, power spots on the earth, shamanistic practices, the identification of the earth as a woman, and the unique abilities of women to tune into the rhythms of nature and tap into the power of the earth because of the nature of women's bodies.

What links this version of the goddess movement with conservatism is the concern with power. It focuses on the strength of women and on what the strength of women has to contribute to the Strict Father family. It does not in any way challenge the Strict Father model as it applies to men, the family, and social arrangements. It accepts the social arrangement in which women do most of the child-rearing and housework and may have to hold down part-time jobs to make ends meet. It addresses the question of how women can find the strength for all this, as well as, quite often, trying to help their husbands overcome such common male tendencies as alcoholism, abusiveness, uncommunicativeness, and inability to express feelings. It also addresses the question of how women can find the strength to deal with the realities of divorce.

To find strength, women look to their bodies, to their emotionality, to their connection to the earth, and to their spiritual power. Women find and hold ceremonies at power spots in the terrain, places where they can tap into the power of the earth. Women lead healing rituals out in nature, rituals in which they sing, beat drums, dance, and allow the power of the body to emerge—rituals that in many ways are like evangelical Protestant church services. Women bring their husbands and families to such rituals so that they can share in the benefits of what women are seen to be best at and come to respect women for the kind of power they have.

All of this can fit quite nicely with most of conservative politics and Strict Father morality. It says women should find the spiritual strength to be as self-reliant as possible. It accepts the moral order in which men have legitimate authority in the family and the public world, but asserts that women have a vital role to play within this structure of authority; and that they have special strengths, nurturant strengths, to be cultivated and respected, even though they are subordinated to the authority of their husbands in a strict father family. Wives should not be helpless; they and their husbands and their children need all the power women can muster.

Women in the conservative goddess movement are conservatives. The earth as a source of power is seen as a resource, both physical and spiritual, for human beings. They do not support liberal ecological ideas such as self-sustainability and environmental protection. They do not support affirmative action. They do not believe that Nurturant Parent morality should apply to the world. They are conservative women, and want respect for being conservative women.

The study of varieties of feminism further confirms the overall theme of this book—that Nurturant Parent and Strict Father models underlie liberalism and conservatism. Conservative forms of feminism may be feminist in various ways, in promoting the idea that women are (or should be) free, strong, competent, responsible for themselves, and deserving of equal opportunity. But conservative forms of feminism do not bring with them Nurturant Parent morality.

Theoretically, the study of varieties of feminism confirms what the study of varieties of liberalism and conservatism confirms, that radial categories are natural and that they arise spontaneously because of natural parameters of variation.

Summary

"Liberal" and "conservative" are not just political categories. They are categories whose central members are defined by family-based moral systems that are projected by the Nation As Family metaphor onto the domain of politics. The categories are then extended in the way categories usually are—by variations on the central models that define noncentral subcategories. The parameters of variation include: (1) Linear Scales; (2) the Pragmatic-Idealistic dimension; (3) Moral Focus; and (4) Moral Order variation.

The nature of such variations is just what research in cognitive science would lead one to expect. The nature of variations within each model reflects the structure of the model.

18

Pathologies, Stereotypes, and Distortions

We saw in the last chapter that there are many variations on the central models of liberalism and conservatism. Not all those variations are welcome to adherents of the central models. Some variations are seen as "pathological." I will use the term "pathological" for *a variation on a central model that subverts the purposes of the central model.* Here is an example.

I observed in Chapter 16 that conservative vigilantism—in which anti-government or anti-liberal violence can be seen as moral action—is a form of conservatism that differs in three ways, all matters of degree, from mainline conservatism. But just because it is a variant on conservatism does not mean that mainline conservatives like it, or find it moral. Indeed, most mainline conservatives work within the law and do not condone breaking the law. To those conservatives, conservative vigilantism is pathological—breaking the law, or preparing for military action against the duly elected government, even in the name of conservative values, violates the conservative principle of obedience to duly constituted authority. Those conservatives disown conservative vigilantes, seeing them as thugs or lunatics.

It is important to look carefully at such "pathological" variations of a central model, variations that subvert the goals of the central moral systems of both liberals and conservatives. It is important, first, simply to be aware of ideological pathologies. Whether you are a conservative or a liberal, you have a responsibility to your moral values. It's important to know when your ideological neighbors are subverting them.

There is a second reason to be aware of pathological variations. The reason is that there is an underhanded practice in politics, as well as in other domains, of what I will call "pathological stereotyping." To see what a pathological stereotype is, let us recall what a stereotype is:

> A social stereotype is a model, widespread in a culture, for making snap judgments—judgments without reflective thought—about an entire category, by virtue of suggesting that the stereotype is the typical case.

A *pathological stereotype* is the use of a pathological variant of a central model to serve as a stereotype for the whole category, and hence to suggest that the pathological variant is typical. Both liberals and conservatives tend to stereotype each other in terms of pathological variations. For example, liberals sometimes stereotype conservatives as "fascists," while conservatives stereotype liberals as "bleeding hearts" or as "permissive." For the sake of fair public discourse, it is important to know just when one side is stereotyping the other in terms of pathological variants and just how those variants differ from the central models.

Pathological Variants of the Family Models

Let us begin with pathological variants of the Nurturant Parent and Strict Father family models. Being a parent is a complex balancing act. If you're a strict parent, you have to balance punishment and reward, you have to set down rules that are strict enough but not too strict, you have to punish severely enough but not too severely, you have to be nurturant but not too nurturant.

The nurturant parent has a similar problem of balance. Nurturing a child requires not just feeding, clothing, and interacting with a child, it also requires guiding the child gradually through loving and respectful interaction to act nurturantly, responsibly, and respectfully toward the rest of the family and the community. It means developing mutually empathetic ways of interacting, so that the child can sense what you want and need. It means communicating expectations and communicating approval and disapproval, explicitly as well as in subtle but effective ways. If all you do is satisfy the child's needs and desires and none of the rest, you're not being nurturant; you're spoiling the child. The problem is that your interactions with the child go one-way only: you respond to his needs and desires and he doesn't respond to yours.

Another variant of one-way interaction is where you simply tell the child what to do and punish him if he doesn't do it. What is going wrong is that you are not nurturing the child, not teaching empathy so that he will know how to behave responsibly in new situations, where empathy, not strict rules, is the guide to appropriate behavior. If you just give and enforce orders, you are not contributing to trust and responsible interdependence. Spoiling and obedience training are two sides of the same coin; both are instances of the same pathology. Let us call this the *Pathology of Insufficiently Nurturant Interaction*. This pathology is defined with respect to the nurturance model, and so we will call it a "nurturance-centered pathology," so that we can keep track of the model in which it is pathological.

The Strict Father model has a corresponding, but very different, pathology. According to the Strict Father model, the way to avoid spoiling a child is to set strict rules and enforce them. You spoil a child by setting insufficiently strict rules or by punishing him insufficiently for breaking the rules. From the perspective of the Strict Father model, what is going wrong is that the parent is being too permissive. Let us call this the *Pathology of Permissiveness*. Since this pathology is defined with respect to the strictness model, let us call it a "strictness-centered pathology."

Since pathologies are defined relative to particular models, they are worldview-dependent. The concept of "permissiveness" only makes sense from the worldview of the Strict Father model, where the job of the parent is to command obedience and enforce it. Correspondingly, the concept of "insufficently nurturant interaction" is defined relative to the Nurturant Parent model.

Because their worldviews are different, each parent would be inherently critical of the other. A strict parent being appropriately strict from his point of view would be insufficiently nurturant from the nurturant parent's viewpoint. A nurturant parent being appropriately nurturant would be permissive from the strict parent's viewpoint.

However, there are certain things they agree on. A child should not be spoiled; that is, just have his needs and desires satisfied. A child needs to grow up to be responsible, self-disciplined, and self-reliant. That is, there is an agreement on certain ends—on certain things that count as successes and failures—though there are radical disagreements about the means and about what is to blame for failures and what is to get credit for success.

POSSIBLE OVERLAPS

The Strict Parent and Nurturant Parent see each other's central models as pathological from their own perspectives. Does this mean that it is impossible to find cases where the values of both models coincide, cases which neither side would find pathological? Such cases are possible and here is one.

Suppose the nurturant parent gently, but explicitly, communicates expectations and gently but effectively expresses approval or disapproval. The expectations communicated could work like rules laid down, and the disapproval would function as punishment. Similarly, the strict parent might engage in nurturant interaction and set the rules gently and clearly in that context and his punishment might amount to effective expression of disapproval. In such a case, one might barely be able to tell one from the other from their behavior, though their logics, their priorities, and their moral impulses would be very different.

Such a near overlap would be quite a stretch for a truly strict parent. It is clearly a noncentral case of the Strict Parent model. Such an overlap could not be maintained indefinitely. Suppose the child challenged the authority of the parent. The strict parent's response is clear: you reestablish authority through discipline, typically corporal punishment, say with a paddle or a thin stick or a belt—not enough to injure the child, but enough to inflict sufficient pain to make him behave. This would be a nurturance-centered pathology.

The nurturant parent's response is very different. You reestablish authority through nurturant communication, in which the parent's ultimate responsibilities are made clear but in which the issues get talked out honestly and thoroughly on both sides and accommodations, if appropriate, are sought. This would be a strictness-centered pathology. Here each parent's response is seen as pathological by the other.

MODEL-INTERNAL PATHOLOGIES

In the above cases, each central model looks pathological to the other. But there is another kind of pathology, where a variant of one's own model seems pathological to you. For example, in the case just discussed, spoiling a child—satisfying its needs and desires and doing nothing else—is seen as an insufficiently nurturant one-way interaction. There is a variant of the nurturance model in which the parent, in the name of nurturance, doesn't really nurture the child but just spoils him. This is pathological from the point of view of the central nurturance model; that is, it is a nurturance-centered pathology. It is, of course, also pathological from the point of view of strictness—a strictness-centered pathology as well.

Another case would be that of an abusive strict parent, one who followed the Strict Father model, except that in administering punishment he really harmed his children. A normal advocate of the Strict Father model would find this variant of his own model pathological. It is a strictness-centered pathology. Of course, it is also a nurturance-centered pathology.

DISTORTIONS OR JUSTIFIED STEREOTYPING?

At this point we can see some common forms of distortion. Let us consider parallel cases.

A nurturant parent stereotypes the strict parent as abusive, suggesting falsely that abuse is characteristic of the central Strict Father model. But the abusive strict parent is just as pathological to the central strict parent as to the nurturant parent.

A strict parent stereotypes the nurturant parent as spoiling, suggesting falsely that spoiling is characteristic of the central nurturant parent model. But the spoiling parent is just as pathological to the central nurturant parent as to the strict parent.

These are clear cases of pathological stereotyping. They are distortions, and such distortions are extremely common, not just in liberal and conservative discourse, but throughout our culture.

There is, however, a perspective from which these are not distortions. The argument that these are not distortions goes as follows: The central models are ideal models, in that they define what an ideal parent would do. But real people are not ideal. They slip up, and the question is, in what direction are they going to slip up? Will common slipups result in the wholesale use of pathological variants of the central model?

For example, a strict parent may see an imperfect nurturant parent as not having the time or patience or training or thoughtfulness to be properly nurturant; an imperfect nurturant parent who is very busy may well wind up spoiling the child.

Similarly, a nurturant parent may see an imperfect strict parent as not having the restraint, empathy, awareness, or sobriety not to harm the child in punishing him; an imperfect strict parent, especially someone who drinks, may well not have the judgment or restraint to keep from abusing the child.

A strict parent, on such grounds, may see it appropriate and correct to stereotype the nurturant parent as spoiling; similarly, a nurturant parent may see it as appropriate to stereotype the strict parent as abusive.

It is not my purpose to justify such a use of pathological stereotyping. My only purpose is to note that it exists and discuss its logic. In order to use this justification, one would have to demonstrate that slipups are the norm and that, in the typical case, a pathological variant is used despite good intentions. This kind of demonstration is rarely, if ever, given. The reason is clear. Pathological stereotyping is not done for the sake of being fair-minded; it is done for the sake of convincing people, fairly or not.

I began this chapter with a discussion of pathological stereotyping of conservatives by liberals. I would like to conclude with two cases of pathological stereotyping of liberals by conservatives.

Pathological Stereotypes of Liberals

Conservatives commonly characterize liberals in three ways: (1) as lovers of bureaucracy; (2) as defenders of special interests; and (3) as advocating only rights and no responsibilities (the "permissive society"). From the liberal perspective, all three are what I have called pathological stereotypes.

First, bureaucracy: Liberals point out that they didn't invent bureaucracy; they inherited it. Bureaucracy was originally brought into government to replace a system of uncontrolled corruption, political favoritism, and unconstrained lobbying. But the dark aspects of bureaucracy—impersonality and unthinking application of regulations—are as antithetical to liberalism as to conservatism. Neither is part of a "nurturant government." Liberals argue further that President Clinton's "Reinventing Government" initiative will change the government in the direction of greatly limiting the bureaucratic dark side.

Second, special interests: Liberals point out that what are called "special interests" by conservatives are, in liberal moral and political philosophy, special cases of serving the public interest, namely, through promoting fairness and other liberal ideals. What conservatives call special interests include groups seeking fair treatment, such as blacks, women, and ethnic minorities. Liberals further point out that conservatives have just as many "special interests" as a result of their philosophy—industries seeking tax breaks, an easing of environmental restrictions, and so on. Real cases of special interests—people seeking favors for themselves that have nothing to do with liberal moral and political philosophy—are anathema to liberals and violate the liberal commitment to social responsibility.

Third, all-rights-and-no-responsibilities: Liberals point out that this is pathological in liberal philosophy. As we have seen, Nurturant Parent morality requires that rights and responsibilities go together. Any variant of liberalism that does not include responsibilities subverts the central model. Most liberals are very concerned with responsibilities. Liberals especially point to the large number of idealistic liberal groups and individuals who have been working for decades for the sake of social responsibility, not on behalf of their own rights.

Pathological Stereotypes of Sixties Liberals

Conservatives—and many people in the media—often represent the sixties generation using three stereotypes:

1. Eternal flower children: hopelessly naive idealists who did nothing but place flowers in their hair, make peace signs, and put into practice the slogan "Make Love, Not War."
2. Deadheads: hedonists interested in nothing but sex, drugs, and rock'n'roll.
3. Violent radicals: loudmouths spouting communist slogans, leading violent antigovernment rallies, and preaching violent revolution.

To members of the sixties generation, however, these are pathological stereotypes, perpetuated by the news media because they make good copy and by conservatives because they serve conservative interests. From the perspective of most sixties liberals, all these were and still are pathological.

Sixties liberalism, to those involved in it, focused on social responsibility. Sixties liberals see it this way: They risked their safety and their lives (and sometimes lost them) in civil rights demonstrations. They fought for anti-poverty programs, and worked to establish them. They brought feminism and ecology into the mainstream. And they demonstrated courageously against what they saw as the immorality, duplicity, and downright foolishness of the federal government in conducting an immoral and ill-advised war in Vietnam. The people who did all these things were not flower children, or deadheads, or violent radicals. They were idealistic liberals who were self-disciplined, self-reliant, hardworking, and dedicated to American ideals; they were the very reverse of the stereotypes. To the typical but unheralded sixties liberals, the eternal flower children, deadheads, and violent radicals were pathological, just as they were to conservatives. That is why, to most sixties liberals, the use of those three stereotypes is pathological stereotyping.

As we have seen in the previous chapters, there was something that united all these disparate themes of sixties liberalism—working for civil rights, against poverty, for feminism, for environmentalism, and against the killing of two to three million citizens of a small third-world country ten thousand miles away. What united them was Nurturant Parent morality applied to politics, the ideal of a nurturant society. That's what made them liberals.

Nurturant Parent morality, as we have seen, includes extremely important notions such as self-nurturance (taking care of oneself), self-development (developing one's potential), doing meaningful work (work that one finds personally fulfilling), and moral happiness (living a basically happy life, which is moral in that one's own joy is a gift to others and in that basic happiness is a prerequisite for genuine empathy).

This was the direction that the sixties generation went in during the seventies. When civil rights legislation had been passed, when feminist and environmental movements and anti-poverty programs had been established, and when the war was over, liberals of the sixties generation turned both to developing themselves as nurturant individuals and to developing a nurturant society.

Concerned with self-development, many turned to the human potential movement and to eastern religions with meditative traditions. The requirement of self-nurturance brought many to a concern with health (physical development and natural foods), healing, and therapy. The concern with moral happiness led to involvement with the aesthetic dimension of life—with art, with living in nature, with sensuality, and with forms of beauty in everyday life. The concern with meaningful work led many into professions like medicine, law, education, and architecture, where they could make a living, work for their social ideals, and practice their personal ideals.

Conservatives, however, have seen all this from their own perspective, a perspective that can only make sense of such people as *self-indulgent, hedonistic yuppies,* selfish people interested only in their own pleasure and advancement. To sixties liberals—those that are "internally-oriented"—this is a pathological stereotype. Within Nurturant Parent morality, taking care of oneself, self-development, and basic happiness (including an aesthetic sense) are moral values functioning in the service of general human well-being. Being self-indulgent and hedonistic is very different from developing one's potential (significantly in the service of others), taking care of oneself (so others don't have to), and developing basic happiness (both in the service of empathy and as a gift for others). To an internally oriented sixties liberal, true self-indulgence and pure hedonism are pathological; they are subversive of social responsibility, which is a central tenet. To sixties liberals, the conservatives' characterization is a pathological stereotype.

Few Are Innocent

Both liberals and conservatives engage regularly in pathological stereotyping. Liberals do it when they characterize conservatives as selfish, abusive fascists, and tools of the rich.

Conservatives do it when they characterize liberals as dedicated to bureaucracy, special interests, and rights-without-responsibilities; and when they stereotype sixties liberals as flower children, deadheads, violent radicals, and self-indulgent hedonists. What pathological stereotyping misses, of course, is morality, the opposing Strict Father and Nurturant Parent moral systems that lie behind central conservatism and liberalism. Pathological stereotyping may serve self-righteousness and propaganda, but it misses all moral understanding.

19

Can There Be a Politics
without Family Values?

Suppose I am right that the divisions in American politics are a reflection of diametrically opposite moral systems based on radically different models of the family. And suppose I am right in my analysis of those moral systems. Then there are important next questions to be asked:

- Can political values in America be separated from family values?
- Is there any way to avoid the application of family values to the political arena?
- Since metaphorical thought lies behind both liberal and conservative politics, is there a way to expel metaphorical thought from politics and from political values?

Can Political Values Be Separated from Family Values?

The nation is not literally a family. The government is not literally our parent. In real families there are genetic bonds and bonds of love between parents and children. No such bonds exist between the government and its citizens. Governments are not even people. But the people who run governments have power—life-and-death power over individual citizens.

Democratic institutions have evolved to protect citizens from the abuse of such power. The separation of powers can be seen from our perspective as a way to keep the government from functioning as an authoritarian strict father. The separation of church and state can be seen as an attempt to insulate our political institutions from the Strict Father morality of many religions.

But morality is by no means absent from democracy. On the contrary. American democratic institutions are based on certain moral schemes, in particular, Moral Fairness, Moral Empathy, and Moral Self-Interest—the maximization of the self-interest of all. Let us begin with Moral Fairness. Among the forms of fairness that are institutionalized are:

- Equality
- Impartial Rule-Based Distribution
- Rights-Based Fairness
- Contractual Distribution

Equality shows up in the one-man, one-vote electoral laws, and in proportional representation in the House of Representatives. Impartial rule-based distribution is the basis of the impartial application of laws. Rights-based fairness is realized in constitutional rights, which the government has a duty to protect. And contractual distribution is realized in the enforcement of contracts.

Moral Self-Interest allows individuals to define what is in their interest in any way they want, with life, liberty, and safety taken for granted as being in one's interest. Moral Self-Interest presupposes Moral Fairness in the form of fairness of competition.

METAPHOR AND THEORETICAL LIBERALISM

Though this book is about political liberalism, not modern theoretical liberalism, it is useful to see how modern theoretical liberalism makes implicit conceptual use of the metaphors we have been considering. Let us consider, in very schematic form, John Rawls's theory of justice as fairness (see References, C3).

Both Moral Fairness and Moral Self-Interest, as they apply in the setting-up of a democratic state, presuppose Moral Empathy. To see why, consider the question ''Why would you want the state that you are born into, or that you enter, to be one that is based on Moral Fairness and Moral Self-Interest (which is 'moral' because it supposedly maximizes the self-interest of everyone)?''

Considering that there are always going to be people who are, in some ways, more powerful than you, you would not want that power differential to result in your being treated unfairly or in having your pursuit of your self-interest squelched. In short, you would want other people to treat you the way they would want to be treated. It is Moral Empathy—putting oneself in other people's shoes—that leads to having a form of democracy based on Moral Fairness and Moral Self-Interest. And to guarantee that there will in fact be Moral Fairness and Moral Self-Interest, you have to have the right kinds of institutions and take part in their governance.

Thus, even in theoretical liberalism, metaphors for morality from the Nurturant Parent model are at the basis of liberal theory.

MORALITY AND POLITICS

Democracy is commonly defined and studied by scholars in terms of liberal institutions, like an independent judiciary and civilian control of the military, not in terms of metaphorical moral concepts. Yet the institutions must function in terms of such moral concepts if they are to be considered really democratic. A form of government in which there are so-called "democratic" institutions, where those institutions do not serve these metaphorical moral schemes—where there is no real self-governance or fairness or moral self-interest— is a hollow democracy at best. Democracy with hollow institutions, institutions that do not realize such moral ideas, is not something worth calling a democracy.

In short, any real democracy comes with some form of morality built into it. Is this form of morality really separable from forms of morality based on models of the family? If the answer were yes, one might argue that any form of morality based on ideal family models ought to be kept out of a democracy.

This is, however, impossible. There are two ways to see why. First, look at the word "ought" in "ought to be kept out of a democracy." What moral principles govern the use of that "ought"? For those whose primary moral principles come from some family-based moral model—either a Strict Father or Nurturant Parent model or perhaps some other, there is no higher set of principles. If you believe that your family-based values are all-encompassing, then you will conclude that those values ought not to be kept out of politics. Because there are people—many people, both liberal and conservative—who do believe that their family values are all-encompassing, those people will never accept some "higher" morality that restricts their family values to non-political arenas.

Second, those family-based values cannot practically be kept out of politics anyway. To see why, consider how Moral Fairness is to apply to Moral Self-Interest to guarantee fair and unrestrained competition. Conservatives and liberals have two different answers depending on their family values. Liberals are concerned with those disadvantaged initially in competition and will want the state to guarantee fairness by helping them. Conservatives will reply that that is coddling—it is "bleeding heart liberalism" that supports moral weakness; instead, they will want to avoid all government "interference" in competition.

There is no neutral answer here. Non-family-based morality—if there is such a thing—just doesn't cover the hard cases, which are most cases. It is family-based morality that gives case-by-case answers in a detailed enough fashion so that one can have policy goals.

Suppose we wanted to separate family-based morality from politics. Suppose we all wanted to have the state set up according to general principles of abstract non-family-based "democratic morality," and that we wanted to keep family-based morality nonpolitical. That would be impossible to carry out in practice. The Nation As Family metaphor will carry family values over into politics whenever those family values are seen as politically relevant. There is no "higher" generally accepted moral or political principle to exclude family values from politics, and there is as well no practical way to exclude them.

Is There a Metaphor-Free Conception of Government?

If one is disturbed by the use of the Nation As Family metaphor, whether in conservative or in liberal discourse, one might ask whether there are alternative, non-family-based metaphors for politics, or even whether it is possible to have a metaphor-free conception of government.

The government is an organization. What kind of an organization is it, or should it be? There is a very long answer to this question, but the short answer is this: Governments have armies and judicial systems, and so governments have in part been modeled metaphorically as armies or as judicial systems. Thus, the American government has a top-to-bottom chain of command, as in an army. It also has a judicial structure in place, with clerks and administrators serving as judges deciding claims brought by citizens according to existing law. In addition, the American government is conceptualized as a business, which is to be run efficiently and not lose money.

In recent years, questions have arisen as to what kind of business it should be. Twentieth-century American bureaucracy was based on an industrial business, a kind of factory model, with bureaucrats as factory managers. This was seen as a way to reform the previous government, which ran largely on political patronage and favoritism. The industrial bureaucracy model was instituted in the late nineteenth and early twentieth century as a reform. To minimize corruption, a system of rules and a staff of civil servants were put in place so as to make government as impartial and efficient as possible. At the time, factories were seen as models of efficiency. Impersonality was seen as a virtue, replacing favoritism and corruption.

The breakdown of the industrial model of governmental organization has been written about at great length in the literature on reinventing government, especially *Reinventing Government* by David Osborne and Ted Gaebler and *Breaking Through Bureaucracy* by Michael Barzelay. The present wisdom has been to replace the industrial metaphor, keeping the Government As Business metaphor, but making it a different kind of business, one specializing in customer service. Taxes, from this perspective, are seen as payments for services rendered to the public, and the impersonality of the factory-like bureaucracy is to be replaced by a more personal form of service.

The government is seen from this perspective as just selling its services to the public for tax money. According to this view, there is no morality in government, just services for sale. When government is framed in that way, it would seem not to have a moral function. It then becomes a practical, not a moral, question as to whether a particular government agency works better than private enterprise. Government as a service industry becomes subject to cost-benefit analysis. Under this model, if the private sector can do a better job, then it should.

Let us take as examples two very different cases. First, take Michael Barzelay's example (1992) of the state of Minnesota motor pool. This is a perfect example of the government-as-service-industry concept. The job of the pool is to provide cars to state officials for the performance of state functions. There is no issue of morality here, just one of efficient operation. If the state motor pool cannot provide better and cheaper service than Hertz and Avis, then it should go out of business. So far, so good.

But compare this with the Environmental Protection Agency. The EPA has not just a practical mission but a moral mission—safeguarding the environment, which includes choosing a moral view of the environment. There is no neutral view of the environment; there are only moral views of the sort we discussed in Chapter 12. The EPA's job is not merely to carry out morally neutral functions like measuring air pollution. Its very function is a moral one. Its regulations, its forms of testing, its research projects, and its sanctions all come out of a moral vision. Parts of its job could be farmed out to the private sector, but its overall job could not, because the market does not incorporate inherent values, such as the inherent value of nature that emerges from the Nurturant Parent model. It is at points like this that family-based morality enters crucially into government.

Many parts of government have functions that are not morally neutral. Those functions cannot be done just as well by private industry, where it is the bottom line, not morality, that matters. For all such cases, government is not just involved in the sale of services for tax money. Instead, the mission of the agency is moral, and its success must be judged in significant part on moral grounds, not cost-benefit grounds. It is the moral mission of the EPA that offends conservatives. The same is true of the moral missions of the arts and humanities endowments.

One of the reasons that conservatives are offended by large segments of the federal government is that those aspects of government have a moral function that does not square with their morality. Take public schools, for example. Our public schools have been shaped by a moral function. They don't just teach the three R's. They teach how to understand our moral life, our history, our politics, and our culture. Public schools have been seen as having a moral mission: to create informed, open-minded, questioning citizens. Their job is not just to teach what is officially sanctioned—say, some officially sanctioned form of U.S. history that hides all the dark and controversial sides of our history. The most crucial part of the job of public schools is to produce independent, informed, questioning citizens. That has been seen as the most important part of their moral mission.

This mission is not independent of family-based morality. Conservatives disagree with that moral mission. It contradicts Strict Father morality. What is an "open" history to liberal educators is a "negative" history to conservatives.

An "open" history that discusses "negative" episodes will include criticisms of the functioning of all forms of morality in American history, including the functioning of Strict Father morality. For example, it might include earlier and harsher versions of the Strict Father family in which children were property who could be sold into indentured servitude or put to work in factories at an early age, and in which wives were seen as chattel. It would certainly include an account of the women's suffrage movement and how it was fought by advocates of traditional Strict Father family life.

By the Principle of Self-Defense, Strict Father morality must be defended at all costs. It must not be subject to criticism in the schools. That would be immoral from its perspective. Hence, any history that puts it in a bad light is "negative."

In addition, many conservatives have a version of the Moral Order metaphor in which the U.S. ranks higher in the Moral Order than any other nation, and so has more moral authority than any other nation. An "open" or "negative" history of the U.S. shows that the U.S. has not acted morally at all times, and that there is much about the morality of our country that could stand improvement. From a liberal perspective, it is good for our country and for our children to think about how the country might be improved morally. But such "benefits" of an "open" history cannot be seen as benefits by those who see the U.S. at the top of the Moral Order of nations. From that perspective, anything "negative," anything that presents less than a pristine image, questions the legitimacy of American moral authority.

There is still another reason for conservatives to be against what they see as a "negative" history. If conservative politics rests on Strict Father morality, the national family must be seen as a moral family and the rules by which the national family operates must be seen as moral. Otherwise, the legitimacy of all forms of governmental authority is called into question. The very foundation of Strict Father morality is the legitimacy of parental authority. To someone raised with Strict Father morality, a "negative" history might call into question that authority. Strict Father morality cannot tolerate the questioning of legitimate authority by children. Children are supposed to venerate and idolize legitimate authorities, so that they can develop character by following the rules laid down by that authority. "Negative" history, conservatives believe, would lead to questioning authority and would threaten that process.

Of course, none of this holds in Nurturant Parent morality, where openness, questioning, and facing one's dark side are virtues. The virtue of the teaching of history to children not as veneration, but as honest, tough-minded inquiry, is a liberal view. Nurturant Parent morality requires open, honest communication, questioning, and explanations. Conservatives are correct that an open/negative U.S. history curriculum fits a liberal moral vision. Liberals are correct that a goody-goody U.S. history is not only inaccurate but fits a conservative moral vision.

Conservatives have called for "a history that is acceptable to all Americans." That means it must be acceptable to conservatives, which in turn means it cannot contain anything that either (1) questions Strict Father morality itself, or (2) questions the idea that the U.S. has more moral authority than any other nation, or (3) questions the moral legitimacy of the U.S. government.

There seems to be no way around it. American politics is suffused with family-based morality. When it comes to specifying policy goals, family-based morality is going to enter—in a big way. Family values are going to matter. The question is, which family values?

Part Six

Who's Right? And How Can You Tell?

20

Nonideological Reasons for Being a Liberal

Up to this point, my aim has been to describe, as accurately as possible, the conceptual systems behind conservatism and liberalism, and to meet as well as possible the standard kinds of criteria for accuracy and explanation that I set for myself initially.

I have tried to put aside my own political views in doing the analysis; setting criteria for an adequate explanation was a way of forcing myself to do that. I hope that, in meeting the criteria for adequacy, I have provided an explanatory model that is worthy of my discipline, a model that is free of political and moral assumptions.

But I am not a moral relativist. I am a committed liberal. In the process of writing this book, I have had to examine, and therefore question, every point of my own beliefs. Every day, I have had to compare my liberal beliefs with conservative beliefs and ask myself what, if any, reason I had to hold my beliefs.

I have emerged from the process with a great respect for the coherence of the conservative position and for the intelligence and cleverness used by conservatives in articulating their views in a powerful way. Like many other liberals, I once thought of conservatives disparagingly as mean, or insensitive, or selfish, or tools of the rich, or just downright fascists. I have come to realize that conservatives are, for the most part, ordinary people who see themselves as highly moral idealists defending what they deeply believe is right. I now understand why there are so many fervently committed conservatives.

I also find conservatism, now that I think I understand it reasonably well, even more frightening than I did before. My new understanding of conservatism and liberalism has made me more of a liberal than ever. I find that I now can consciously comprehend my old instincts. I can give names to things that I could not clearly articulate before, things that were part of a vague sense of what was right. What's more important is that I understand that political liberalism comes out of a well-grounded, highly structured, and fully developed moral system that I deeply believe in. That moral system itself comes out of a model of the family that I also deeply believe in. Now that I can see the unity and strength of liberal morality and politics, I feel more than ever that liberalism must be articulated fully, communicated clearly, and defended staunchly, not on an issue-by-issue basis, but as a whole, as a deeply moral perspective on politics.

It may sound as if my old prejudices have just been reinforced and that I am taking liberalism on faith alone. Not so. The process of thinking all this through has convinced me that there are overwhelming reasons to be a liberal that come from outside liberalism itself. I think I can finally put my finger on just why I have been a liberal, why political liberalism has always made sense to me, why it has always been not only an idealistic and practical calling but mainly a response to my most fundamental human instincts. What I have found, in the course of this study, is that there are in fact good reasons for choosing the Nurturant Parent model of the family, Nurturant Parent morality, and with them, liberal politics.

This book has been about worldview. But at some point the world must enter. Liberals and conservatives have different ideal models of how to raise children. Are there reasons to choose? Liberals and conservatives have different moral systems. Does it make any sense to compare moral systems? If so, on what basis can you compare them, and what results from the comparison? Liberals and conservatives make different assumptions about how people naturally think and act. Do we know enough from cognitive (or any other) science to decide the matter?

I believe that we know enough to allow us to choose in all these domains. It is our knowledge of the world that allows us to choose between worldviews. In each case, there is research bearing on the choice and in each case the answer is the same: There are good reasons to choose liberalism.

If I were to be asked to list those reasons and the bases for them very briefly at the outset, each in one or two sentences, here is what that list would be:

> Reason 1. The Nurturant Parent model is superior as a method of childrearing.
>
> Reason 2. Strict Father morality requires a view of human thought that is at odds with what we know about the way the mind works.
>
> Reason 3. Strict Father morality often finds morality in harm; Nurturant Parent morality does not.

There are, of course, other real-world reasons to be a liberal. The environment is being seriously threatened right now, and conservative moves to end environmental controls, such as clean-air and pure-water enforcement, will only make things much worse. At present, 70 percent of the wealth in America is owned by 10 percent of the families. That means that 90 percent of the families share only 30 percent of the wealth. Since the rich have always tended to get richer, not poorer, the prospects are for a less than 30 percent share of the national wealth to be available to 90 percent of our population. That disparity is so large that it threatens the possibility for real prosperity for most of our citizens. Further conservative tax cuts for the rich will only make the disparity larger.

These and many other real-world reasons make conservatism dangerous. But there is no lack of competent observers writing about these issues. Let me turn instead to the three reasons I just mentioned, since the public is not widely aware of them.

21

Raising Real Children

The conservative family values agenda is, at present, being set primarily by fundamentalist Christians. This is not a situation that many people are aware of. Probably the most prominent figures in the fundamentalist Christian family values movement are Dr. James Dobson, who is president of Focus on the Family, based in Colorado Springs, and Gary L. Bauer, who runs the Family Research Council in Washington, D.C. These groups have been most explicit in developing a Strict Father approach to childrearing and have been extremely active in promoting their approach. On the whole, they are defining the conservative position for the current debate about childrearing, as well as for legislation incorporating their approach. Since the ideas in conservative Christian childrearing manuals are fully consistent with the Strict Father model of the family that lies behind conservative politics, it is not at all strange that such fundamentalist groups should be setting the national conservative agenda on family values.

I should say at the outset that virtually all of the mainstream experts on childrearing see the Strict Father model as being destructive to children. A nurturant approach is preferred. And most of the child development literature within the field of developmental psychology points in one direction: childrearing according to the Strict Father model harms children; a Nurturant Parent model is far superior.

In short, conservative family values, which are the basis for conservative morality and political thought, are not supported by either research in child development or the mainstream childrearing experts in the country. That is another reason why the conservative family agenda has been left to fundamentalist Christians. Since there is no significant body of mainstream experts who support the Strict Father model, conservatives can rely only on fundamentalist Christians, who have the only well thought out approach to childrearing that supports the Strict Family model.

The claims to legitimacy for the conservative family values enterprise rest with the fundamentalist Christian community, a community whose conclusions are not based on empirical research but on a fundamentalist interpretation of the Bible. And that, as was shown in Chapter 14, is based on Strict Father morality itself. Thus, there is no independent or nonideological basis whatever for conservative claims about family values.

The conservative Christians who set the conservative family values agenda are not particularly interested in empirical research or the wisdom of the extensive community of mainstream experts on childrearing. As James Dobson puts it,

> I don't believe the scientific community is the best source of information on proper parenting techniques. There have been some worthwhile studies to be sure. But the subject of parent-child interaction is incredibly complex and subtle. The only way to investigate it scientifically is to reduce the relationship to its simplest common denominators, so it can be examined.

But in doing so, the overall tone is missed. Some things in life are so complicated that they defy rigorous scrutiny, and parental discipline (in my view) appears to be one of them.

The best source of guidance for parents can be found in the wisdom of the Judeo-Christian ethic, which originated with the Creator and has been handed down generation by generation from the time of Christ. (B3, Dobson, *The New Dare to Discipline*, p. 16)

I simply do not agree that research about childrearing is irrelevant. There are important things to know. What are the effects of punishing children, especially beating them with sticks, belts, and paddles? Are there physical effects? Long-term psychological effects? Is there any correlation between punishment by beating and humiliation and violent behavior later in life? Do most delinquent children have a history of strict parenting, nurturant parenting, or is it fifty-fifty? What is the effect of first whipping a child and then hugging her? What is the effect of breaking down a child's will by hitting her with a stick? What is the effect of demanding absolute obedience to a father's authority?

To see more clearly what is at stake in knowing about research on such matters, let us look closely at what some conservative Christian childrearing manuals have to say about how children should be raised. These manuals are clear on many points:

1. Children are inherently sinful and defiant.
2. Only punishment and reward will train children away from defiance and pursuing their sinful desires.
3. The only way a child can be raised properly is for a father to demand absolute obedience to his authority. Any questioning of authority requires swift and painful punishment.

4. Obedience can be taught only through painful corporal punishment—by whipping with belts or beating with switches or paddles.

5. Continued disobedience requires greater beating.

6. Punishment for disobedience is a form of love.

7. Parental authority is a proper model for all authority, and children must learn to obey authority so that they can wield it properly in later life.

The following quotations are taken from Dr. James Dobson, J. Richard Fugate, Reverend Jack Hyles, Larry Christenson, and Larry Tomczak (References, B3). Dobson, as you will see, is the most moderate figure. The others are more extreme.

Dobson discusses behaviorist (reward and punishment) principles of child rearing at great length. Though his main focus is on punishment, he also suggests rewards:

> Everything worth having comes with a price. (Dobson, 126)
>
> Two pennies should be granted for every behavior done properly in a given day. If more than three items are missed, no pennies should be granted. (Dobson, 85)

But Dobson is clear about the need for punishment, as are the others.

> Rewards should not be used as a substitute for authority; reward and punishment each has its proper place in child management, and reversals bring unfortunate results. (Dobson, 91)

The point of punishment is not for some specific offense, but to enforce the parent's absolute authority in general, as a matter of principle. Any rebelliousness of spirit must be broken.

> When youngsters display stiff-necked rebellion, you must be willing to respond to the challenge immedi-

ately. When nose-to-nose confrontation occurs between you and your child, it is not the time to discuss the virtues of obedience. It is not the occasion to send him to his room to pout. Nor is it the time to postpone disciplinary measures till your tired spouse plods home from work.

You have drawn a line in the dirt, and the child has deliberately flopped his bony little toe across it. Who is going to win? Who has the most courage? (Dobson, 20)

The only issue in rebellion is will; in other words, who is going to rule, the parent or the child. The major objective of chastisement [that is, physical punishment] is forcing the child's obedience to the will of his parents. (Fugate, 143)

The spanking should be administered firmly. It should be painful and it should last until the child's will is broken. It should last until the child is crying, not tears of anger, but tears of a broken will. As long as he is stiff, grits his teeth, holds on to his own will, the spanking should continue. (Hyles, 99–100)

In the [biblical] command of obedience given to children, there is no mention made of any exception. It must be set forth and impressed on them without any exception. "But what if parents command something wrong?" This is precocious inquisitiveness. Such a question should perish on the lips of a Christian child. (Christenson, 59)

Require strict obedience. The obedience should always be immediate, instant, without question or argument. What the father says to do, the son does. He does it well, he does it immediately, he does it without argument. The parents allow no exceptions to the rule. Hence, obedience is the law of the land and the child should not deem it necessary to have an explanation for orders he has received from his parents. (Hyles, 144)

Obedience is the most necessary ingredient to be required from the child. This is especially true for a girl, for she must be obedient all her life. The boy who is obedient to his mother and father will some day become the head of the home; not so for the girl. Whereas the boy is being trained to be a leader, the girl is being trained to be a follower. Hence, obedience is far more important to her, for she must some day transfer it from her parents to her husband. . . . This means that she should never be allowed to argue at all. She should become submissive and obedient. She must obey immediately, without question, and without argument. The parents who require this have done a big favor for their future son-in-law. (Hyles, 158)

Swift and painful punishment is thus seen as the basis for all character development:

Obedience is the foundation for all character. It is the foundation for the home. It is the foundation for a school. It is the foundation for a society. It is absolutely necessary for law and order to prevail. (Hyles, 145)

The means of punishment is also generally agreed upon. The "rod" in "Spare the rod and spoil the child" is meant literally:

The Biblical definition of the rod is a small flexible branch from a tree (a wooden stick) . . . a number of rods [should be kept] throughout the house, in your

car, and in your purse [so that you can] apply loving
correction immediately. (Tomczak, 117)

The rod is to be a thick wooden stick like a switch.
Of course, the size of the rod should vary with the
size of the child. A willow or peach tree branch may
be fine for a rebellious two-year-old, but a small hick-
ory rod or dowel rod would be more fitting for a
well-muscled teenage boy. (Fugate, 141)

The use of the rod enables a controlled administration
of pain to obtain submission and future obedience. If
a child's rebellion has been to disobey an instruction
willfully, the parent can stop after a sufficient number
of strokes and ask the child if he will obey instruc-
tions in the future. The parent is the best judge of the
correct number and intensity of strokes needed for a
particular child. However, if the child repeatedly dis-
obeys, the chastisement has not been painful enough.
(Fugate, 142–43)

Since such punishment is necessary to form character, it is
a form of love.

Disciplinary action is not an assault on parental love;
it is a function of it. Appropriate punishment is not
something parents do *to* a beloved child; it is something done *for* him or her. (Dobson, 22)

Because I love you so much, I must teach you to
obey me. (Dobson, 55)

When the child is grown up, he must be sent off on his own.
Any parental protection would be harmful:

Unfortunately, many North American parents still
"bail out" their children long after they are grown

346 • CHAPTER TWENTY-ONE

and living away from home. What is the result? This
overprotection produces emotional cripples who often
develop lasting characteristics of dependency and a
kind of perpetual adolescence. (Dobson, 116)

Dobson, like other writers, is also clear about what not to
do: any form of child rearing that does not use painful punishment to enforce absolute obedience to parental authority is "permissive" and promotes "self-indulgence." When
conservatives speak about permissiveness, this is what they
mean.

How inaccurate is the belief that self-control is maximized in an environment that places no obligations
on its children. How foolish is the assumption that
self-discipline is a product of self-indulgence. (Dobson, 173)

Incidentally, Dobson is one of the less extreme conservatives. And despite his disdain for scientific research, he
incorporates some of it into his teachings. Here are some
examples where he uses what has been learned in child
development research:

There is no excuse for spanking babies younger than
fifteen or eighteen months of age. (Dobson, 65)

Parents cannot require their children to treat them
with dignity if they will not do the same in return.
Parents should be gentle with their child's ego, never
belittling or embarrassing him in front of friends. . . .
Self-esteem is the valuable attribute in human nature.
It can be damaged by very minor incidents, and its reconstruction is often difficult to engineer. Thus, a father who is sarcastic and biting in his criticism of children cannot expect to receive genuine respect in
return. (Dobson, 25–26)

Although Dobson does not mention attachment theory (which we will discuss shortly) by name or cite any references, he is obviously aware of the literature on the subject:

> Parents who are cold and stern with their sons and daughters often leave them damaged for life. (Dobson, 12)

> In homes where children are not adored by at least one parent (or a parent-figure), they wither like a plant without water. (Dobson, 48)

> Hundreds of more recent studies indicate that the mother-child relationship during the first year of life is apparently vital to the infant's survival. An unloved child is truly the saddest phenomenon in all nature. (Dobson, 49)

Interestingly, Dobson here is only citing the mother-child studies, not the father-child studies that show that fathers can develop just as effective secure attachments as mothers can. This important omission fits in with Dobson's view that the father is the proper head of the family and the mother's job is to stay home and raise the children.

In addition, Dobson assumes, contrary to attachment theory, that unconditional love "spoils" a child:

> While the absence of love has a predictable effect on children, it is not so well known that excessive love or "super love" imposes its hazards too. I believe that some children are spoiled by love. (Dobson, 49)

These occasional nods to research results are, however, not the main message that Dobson is getting across. Such passages occur briefly and only occasionally and go against the main flow of what the books have to say. The bulk of Dobson's books are about authority and swift, painful punishment. After all, he called his classic book *Dare to Discipline,* not *The Fragile Ego of the Child* or *Don't Spank Your Baby*. Dobson uses research results not to reevaluate his general claims, which come from his interpretation of the Bible, but only to rein in some of the most obviously dangerous impulses of strict fathers. Yet in the overall context of his work, such passages tend to get lost.

We can now see a bit better what is meant when members of the conservative family values movement talk about "discipline," "parental authority," "spanking," and "traditional family values." "Spanking" means hitting a child, starting in toddlerhood, with a belt, a paddle, or the branch of a tree.

The conservative family values movement is pushing hard to stop the funding of social workers who investigate child abuse. They especially want evidence from bruises incurred during "spanking" not to count as evidence of child abuse:

> social workers seeking to rescue children from abusive homes often have . . . problems being fair. Many good parents in loving homes have lost custody of their sons and daughters because of evidence that is misinterpreted. For example, a dime-sized bruise on the buttocks of a fair-skinned child may or may not indicate an abusive situation. It all depends. In an otherwise secure and loving home, that bruise may have no greater psychological impact than a skinned knee or stubbed toe. (Dobson, p. 25)

Gary L. Bauer's Family Research Council has been crusading against all efforts to ban the corporal punishment of children. It has also been trying to get funds taken away from child protective services such as social workers investigating child abuse. Bauer sees such investigations as invasions of privacy by the "therapeutic sector."

Is this group of fundamentalist Christians representative of conservative attitudes about childrearing? I don't know, but they are in charge. They are the people setting the conservative family values agenda.

There is no lack of research on the effects of such Strict Father parenting. Indeed, there is a lot of it. It is not possible for me here to survey anywhere near all of it. That would require a book much longer than this one. However, I would like to give the reader a sense of what some of that research is and the direction in which it points.

Attachment Theory

What is it that leads to disturbed family relationships, to child abuse, to alienated, dysfunctional adults who have little stake in society? There are many lines of research into this question. One of the principal ones is attachment theory. It was first developed by John Bowlby and Mary Ainsworth thirty years ago, and has now become a mature, well-respected, and far-flung research endeavor. For an excellent popular survey of this research, see *Becoming Attached,* by Robert Karen (see References, B1, for introductory material). The final answers are not in, but here is what attachment theory indicates at present.

Attachment theory over the past thirty years has documented the disastrous effects of

> the old-fashioned . . . style of parenting, which was
> impatient with the child's emotional demands, which
> held that the greatest sin was to spoil children by
> showing too much concern for their outbursts, pro-
> tests, or plaints, which was insensitive to the harm
> done by separating the child from its primary care-
> giver, and which held that strict discipline was
> the surest route to maturity. (Karen, p. 50)

Attachment theory indicates the opposite, that "getting love reliably and consistently makes the child feel worthy of love;

and his perception that he can attain what he needs from those around him yields the sense that he is an effective person who can have an impact on his world'' (Karen, p. 242).

Self-discipline and self-denial are not what makes children self-reliant. Nurturance does not spoil children. As Mary Ainsworth says, ''It's a good thing to give a baby and a young child physical contact, especially when they want it and seek it. It doesn't spoil them. It doesn't make them clingy. It doesn't make them addicted to being held'' (Karen, p. 173). This is supported by longitudinal studies. ''Babies cried less at twelve months if their cries had been responded to conscientiously when they were younger'' (Karen, p. 173). ''Whatever relationship advantages secure attachment does tend to confer persist through age fifteen'' (Karen, p. 202), which is as long as the studies have been carried out. The latter is a remarkable finding; secure attachments developed early have a lasting effect.

The basic claim of attachment theory, considerably oversimplified, is this: A child will function better in later life if he is ''securely attached'' to his mother or father or other caregiver from birth. That is, he will be more self-reliant, responsible, socially adept, and confident. Secure attachment arises from regular, loving interaction, especially when the child desires it. Letting a child go it alone and tough it out, denying him loving interaction when he wants it does not create strength, confidence, and self-reliance. It creates ''avoidant attachment''—lack of trust, difficulties in relating positively to others, lack of respect for and responsibility toward others, and in many cases antisocial or criminal behavior and rage. Alternate unsure experiences of attachment and avoidance by parents create a third type of attachment: ambivalent attachment, which results in ambivalent behavior towards others in later life, a dread of abandonment and an inability to see one's own responsibility in relationships, and continuing feelings of anger and hurt toward one's parents. Ambivalent attachment might arise, for example, from painful punishment (to enforce obedience) followed by extreme affection (to show daddy loves you).

These results appear at present to support the values of the Nurturant Parent model over the Strict Father model.

Importantly, it is not just Strict Father family values that harm children. Consider a young, impoverished, uneducated single mother who does not know how to nurture a child properly and who hits or ignores her child when he needs attention. The effect may be avoidant attachment coming from a source other than Strict Father parenting, namely, neglect. In the American context, it is a bit ironic that the Strict Father model applied in a two-parent family may have effects that are similar to those of families with inattentive or violent single mothers, where there is no father, strict or otherwise. The issue is not one parent or two. The issue is the quality of nurturance.

Critiques of attachment theory are varied: Some critics suggest a greater role for genetic predisposition and some suggest that the results are culturally relative. But no major body of research supports the Strict Father model on this issue. So far as present results show, the denial of secure attachment does not build self-reliance and responsibility for others, as advocates of Strict Father parenting imply.

One important critique of attachment theory is that it focuses mainly on early childhood. Yet, as of 1993, the results hold up to the age of fifteen (B1, Sroufe et al. 1992; Karen, p. 202).

Socialization Research

There is other research that has focused on what happens later or throughout childhood. The closest that I have found so far to a head-to-head comparison between the Strict Father and Nurturant Parent models is research in the tradition of Diana Baumrind's fourfold scheme. The best survey I know of this research, though it only goes up to the early 1980s, is in Maccoby and Martin's classic paper, "Socialization in the context of the family: Parent-child interaction," in the fourth edition of the *Handbook of Child Psychology*, edited by Paul Mussen, which appeared in 1983.

Baumrind distinguishes between what she calls "authoritarian" and "authoritative" childrearing styles, the authoritarian being what I have called in more neutral language "the Strict Father" model and the "authoritative" being a version of what I have called "the Nurturant Parent" model. Here are her descriptions of the two models:

The Authoritarian Model

1. Attempting to shape, control, and evaluate the behavior and attitudes of one's children in accordance with an absolute set of standards.
2. Valuing obedience, respect for authority, work, tradition, and preservation of order.
3. Discouraging verbal give-and-take between parent and child.

The Authoritative Model

1. Expectation for mature behavior from child and clear standard setting.
2. Firm enforcement of rules and standards using commands and sanctions when necessary.
3. Encouragement of the child's independence and individuality.
4. Open communication between parents and children, with parents listening to children's point of view, as well as expressing their own; encouragement of verbal give-and-take.
5. Recognition of rights of both parents and children.

"Firm enforcement" and "sanctions" do not include painful corporal punishment.

Catherine Lewis (B2, 1981) has made two important observations about the inclusion of "firm enforcement of rules and standards" in Baumrind's model. The first is a technical point: the way that Baumrind defined "firm enforcement" includes items that reflect success in obtaining obedience, which, Lewis argues, amounts to "low parent-child conflict." Lewis also shows that if the "firm enforcement" part of the model is simply omitted from the pattern of behaviors studied, the results are essentially the same. This indicates that "firm enforcement" does not add anything to the model; in short, the effect of the rest of the model would appear to be that it creates low parent-child conflict and hence the effect of firm enforcement without the need for firm enforcement.

Baumrind's response to Lewis's criticism goes as follows (B2, Baumrind 1991):

> Lewis (1981) has challenged the importance I attach to the *pattern of firm control and high maturity demands*. In her thoughtful critique of my interpretation of the effects of firm control, she suggested that neither demanding practices nor authoritative childrearing is necessary to the development of optimal competence. She is correct. As we have seen, authoritative childrearing was sufficient but not necessary to produce competence and prevent incompetence, as these terms were defined in the study; and demanding practices were sufficient but not necessary to produce social assertiveness in girls. Authoritative childrearing was the only pattern that consistently produced optimally competent children and failed to produce incompetent children in the preschool years and in middle childhood, and this was true for both boys and girls.

Some came from *harmonious* homes. Harmonious parents are highly responsive and moderately firm but attach little importance to obtaining obedience.

Let us now consider the harmonious model.

THE HARMONIOUS MODEL

1. Expectation for mature behavior from child and clear standard setting.
2. High responsiveness, moderate firmness, little importance given to obtaining obedience.
3. Encouragement of the child's independence and individuality.
4. Open communication between parents and children, with parents listening to children's point of view, as well as expressing their own; encouragement of verbal give-and-take.
5. Recognition of rights of both parents and children.

To get some sense of the results of these studies, let us begin by looking at Maccoby and Martin's survey of a wide range of studies of authoritarian childrearing.

> Children of authoritarian parents tend to lack social competence with peers: They tend to withdraw, not to take social initiative, to lack spontaneity. Although they do not behave differently from children of other types of parents on contrived measures of resistance to temptation, on projective tests and parent reports they do show lesser evidence of "conscience" and are more likely to have external, rather than internal, moral orientation in discussing what is the "right" behavior in situations of moral conflict. In boys, there is evidence that motivation for intellectual performance is low. Several studies link authoritarian parenting with low self-esteem and external locus of control.

Whereas the parents of aggressive children tend to be authoritarian, children of authoritarian parents may or may not be aggressive, and so far the aspects of family interaction that are important in determining whether a child of authoritarian parenting will be subdued or "out of control" have not been satisfactorily identified. (B2, Maccoby and Martin, p. 44)

Let's go over this point by point to see what it means in detail.

The Strict Father (or "authoritarian") model is supposed to make a child strong and better able to function socially. It is supposed to make children into effective leaders. But, in fact, it has the opposite effect. Children of the authoritarian parent "lack social competence with peers: They tend to withdraw, not to take social initiative, to lack spontaneity."

Strictly enforced obedience to authority is supposed to make children internally strong and self-disciplined so that they can resist temptations. But it doesn't work. Children of authoritarian parents "do not behave differently from children of other types of parents on contrived measures of resistance to temptation."

An upbringing with strict rules and punishments for violating them is supposed to produce a strong conscience in children. But the opposite is true. Such children show lesser evidence of conscience.

Getting children to follow strict rules through punitive enforcement is supposed to make them morally self-reliant, to create in them an inner moral sense that they can apply to new situations of moral conflict. But again the opposite is true. Such children are more likely to have to depend on the moral opinion of others, that is, they are "more likely to have external, rather than internal, moral orientation in discussing what is the 'right' behavior in situations of moral conflict."

Strict discipline is supposed to make a child internally strong and able to control himself and thus to produce in him a high sense of self-esteem. Again the opposite is true. "Authoritarian parenting" is linked "with low self-esteem and external locus of control," the need for someone else to be in control.

Learning obedience through punishment is supposed to eliminate all aggressive behavior toward parents, to produce respectful behavior toward parents, and to produce nonaggressive, respectful behavior toward others. But that isn't true either. Where do aggressive children tend to come from? "The parents of aggressive children tend to be authoritarian."

In short, the aggressive "out of control" children tend to be the products of authoritarian upbringing. But the converse is not true. An authoritarian upbringing does not always result in aggressive "out of control" children. Sometimes such children are subdued, but it is not yet known what additional factors tend to make them so.

This overall picture is quite damning for the Strict Father model. That model seems to be a myth. If this research is right, a Strict Father upbringing does not produce the kind of child it claims to produce. Incidentally, this picture is not from one study or from studies by one researcher. This is the overall picture gathered from many studies by many different researchers (see References, B2).

And what about the authoritative model, the one like the Nurturant Parent model? What follows is Maccoby and Martin's summary of a wide range of research by many researchers. The results are essentially the same as for the harmonious model.

> The authoritative-reciprocal pattern of parenting is associated with children's being independent, "agentic" in both the cognitive and social spheres, socially responsi-

ble, able to control aggression, self-confident, and high in self-esteem. (B2, Maccoby and Martin, p. 48)

Again, let's look at the details.

The "authoritative" parent, essentially what I have called the "nurturant" parent, encourages independence, originality, and open communication, and listens to the child's point of view as well as expressing his own. The result is not dependence, as the Strict Father model would predict, but independence, just as the Nurturant Parent model does predict.

The Nurturant Parent model predicts that by encouraging independence and engaging the child in dialogue, the child will become "agentic," that is, able to function on his own both mentally and socially. This is the opposite of what the Strict Father model would predict, that only through strict punitive discipline enforcing obedience to an external authority can children internalize authority and be able to think and act on their own. The research shows that this prediction of the Strict Father model is false.

The Nurturant Parent model predicts that encouragement, respect, and being listened to seriously should enable children to be able to exercise self-control, act confidently, and have high self-esteem. The research indicates that such a strategy does work. Again the result is the opposite of what the Strict Father model would predict.

The Nurturant Parent model predicts that if children get to openly discuss reasons for what they are being told to do and how their actions will affect other people, then they will become socially responsible. Again, this is what happens.

In short, the authoritarian (Strict Father) model fails miserably at raising children; the authoritative (Nurturant Parent) model works extremely well.

There is a relatively small difference between the effects of the authoritative and the harmonious models. As Baumrind

(1991) reports, "The children from harmonious families, in comparison with those from authoritative families, were somewhat less assertive than they were socially responsible" (B2, Baumrind 1991, p. 364).

Incidentally, Diana Baumrind's categorization includes two other models, the indulgent-permissive model and the indifferent-uninvolved model. These are the two models that advocates of the Strict Father model usually attribute incorrectly to nurturant parents. Advocates of the Strict Father model tend to make the mistake of lumping together all models of parenting that do not have as their overriding concern unquestioning obedience enforced by painful punishment. They see anything else as neglectful and indulgent. They are not even taking into account the Nurturant Parent model.

Research shows that indulgence and neglect produce what both the Strict Father and Nurturant Parent models would expect. Here are the two models:

THE INDULGENT-PERMISSIVE MODEL

1. Taking a tolerant, accepting attitude toward the child's impulses, including sexual and aggressive impulses.
2. Using little punishment and avoiding, whenever possible, asserting authority or imposing controls or restrictions.
3. Making few demands for mature behavior (e.g., having manners or carrying out tasks).
4. Allowing children to regulate their own behavior and make their own decisions when at all possible.
5. Having few rules governing the child's time-schedule (bedtime, mealtime, TV watching).

THE INDIFFERENT-UNINVOLVED MODEL

1. Tending to orient one's behavior primarily toward the avoidance of inconvenience.

2. Responding to immediate demands from children in such a way as to terminate the demands.
3. Being psychologically unavailable.

Maccoby and Martin summarize the findings for the Indulgent-Permissive model as follows:

> It appears on the whole to have more negative than positive effects, in the sense that it is associated with children's being impulsive, aggressive, and lacking in independence and in the ability to take responsibility. (B2, Maccoby and Martin, pp. 45–46)

The findings for the Indifferent-Uninvolved model were as follows (see Maccoby and Martin, pp. 48–51): Children of psychologically unavailable mothers showed deficits in all aspects of psychological functioning by the age of two, greater deficits than occurred with the other patterns of parental maltreatment. In four- to five-and-a-half-year olds, paternal uninvolvement correlated with aggressiveness and disobedience. Things get worse by the age of fourteen: Children were:

> impulsive (in the sense of lacking in concentration, being moody, spending money quickly rather than saving it, and having difficulty controlling aggressive outbursts), uninterested in school, likely to be truant or spend time on the streets or at discos; in addition, their friends were often disliked by their parents. [They] tended to start drinking, smoking, and heterosexual dating at earlier ages. Continuities to the age of 20 were found. At this age, [they were more likely] to be hedonistic and lack tolerance for frustration and emotional control; they also lacked long-term goals, drank to excess, and more often had a record of arrests.

They were also less likely to have strong achievement motives and to be oriented to the future. Neither of these findings would surprise either an authoritarian, authoritative, or harmonious parent.

But the findings that distinguish the authoritarian from the authoritative models should be extremely disturbing to advocates of the Strict Father model, since they indicate that that model fails thoroughly, though the Indifference-Uninvolved model may be even worse.

Research of this sort is anything but final. No responsible scientist would say, and none does, that any of the above has been absolutely proved. It all needs to be elaborated, extended, checked, and integrated. Other research paradigms need to be developed as well, to provide a cross-check. All this is true of any scientific research. Yet the direction in which the results point is unmistakable. The findings are not random and all over the place. They form a clear pattern and they fit with other findings, which we will look at now.

Obedience, Punishment, and Violence

Let us now turn to the question of the effects of physical punishment on children. This research should also not give solace to advocates of the Strict Father model. The major research indicates that having strict parents who perform painful corporal punishment in childhood leads to domestic violence, aggression, and delinquency in later life.

Take Richard Gelles' *The Violent Home* (B5, 1974), which was based on interviews in a New Hampshire community and which discovered an astonishing amount of physical violence in the households there. Gelles found that "many of the respondents who had committed acts of violence toward their spouses had been exposed to conjugal violence as children and had been frequent victims of parental violence." Gelles became convinced that "the family serves as a basic training ground for violence by exposing children to violence, by making them the victims of violence, and by providing them with the learning contexts for the commission of violent acts. . . . The family inculcates children with normative and value systems that approve of the use of violence on family members in various situations" (Gelles, pp. 58–78, and passim).

These observations are confirmed by Murray Straus, Richard Gelles, and Suzanne Steinmetz in *Behind Closed Doors: Violence in the American Family* (B5, 1981). The authors have found that domestic violence of some sort occurs in half of the households in America. They argue that physical punishments are violent acts that lead to further violence between spouses. Richard J. Gelles and Murray Straus, in *Intimate Violence* (B5, 1988), conclude that "After two decades of research on the causes and consequences of family violence, we are convinced that our society must abandon its reliance on spanking children if we are to prevent intimate violence" (p. 197).

Straus, Gelles, and Steinmetz observe that "The people who experienced the most punishment as teen-agers have a rate of wife-beating and husband-beating that is four times greater than those whose parents did not hit them" (p. 3). Further studies indicating the same are cited in the References, sec. B5. The relationship between these studies and the history of corporal punishment in fundamentalist Protestantism is the subject of an important book by Philip Greven, *Spare the Child: The Religious Roots of Punishment and the Psychological Impact of Physical Abuse.*

As I remarked above, it would take at least a book this size to thoroughly survey all the research bearing on the relative merits of the Strict Father versus the Nurturant Parent models. But so far as I have been able to find out, in the reading I have done and in conversations I have had with professionals doing such research, the results from many different research paradigms point in the same direction: The Strict Father model is bad for children and tends to do the opposite of what it sets out to do. The Nurturant Parent model, on the other hand, works extremely well.

Mainstream Childrearing Manuals

The mainstream books on childrearing, incidentally, reflect all this research. Many are written by highly qualified professional pediatricians who also do research, writing in the tradition of Dr. Benjamin Spock. They have studied child development, kept up on what has been found out recently, cared for real children for years, and have done some of that research themselves. A trip to the parenting section of a good general bookstore will show shelves of books on how to be a nurturant parent—not an indulgent or neglectful parent, but a truly nurturant parent.

A good example is the best-selling *Touchpoints,* by T. Berry Brazelton, M.D., perhaps the country's best known and most respected practicing pediatrician. Unlike the fundamentalist Christian manuals, which pay little if any attention to the stages of child development, the Brazelton book pays great attention to each stage, since any nurturant parent, out of empathy, will want to know as much as possible about what capacities and difficulties the child has at each stage. As Brazelton points out in his section on discipline,

> At each stage of development, there are kinds of behavior that seem too aggressive and out of control, but that are actually normal. If you overreact to them at this exploratory phase, you may end up by reinforcing them. (B4, Brazelton, 255)

Brazelton begins the Discipline chapter as follows, "Next to love, discipline is a parent's second most important gift to a child" (252). He starts the very next section by saying, "*Discipline* means 'teaching,' not punishment." Brazelton gives a list of normal aggressive behaviors for each stage of development. Between eighteen and thirty months (eighteen months is the age at which James Dobson says that painful punitive discipline is to start), Brazelton observes:

> Temper tantrums and violently negative behavior begin to appear at this age. A natural and critical surge of independence comes in the second and third years. The child is trying to separate from you and learn to make her own decisions. . . . It's not possible to avoid tantrums, so don't try. . . . The more involved you are, the longer they will last. It's often wisest simply to make sure she can't hurt herself and just walk out of the room. . . . When she's able to listen, try to let her know that you understand how hard it is to be two or three and to be unable to make up one's own mind. But let her know that she *will* learn how and that, meanwhile, it's okay to lose control.
> (Brazelton, 257)

This is very different than using belts and rods to try to break a child's will. What does Brazelton say about physical punishment?

> *Physical punishment has very real disadvantages.* Remember what it means to a child to see you lose control and act physically aggressive. It means you believe in power and physical aggression. (Brazelton, 260)

This is a wonderfully clear way to say that violence begets violence—and explain why. It's not that Brazelton doesn't believe in discipline. He spends a great deal of time on what discipline is, at each stage of development, and why physical punishment isn't discipline. The important things are knowing what kinds of acting-out are normal, and how to react but not overreact: talking things over with the child so she can understand why she's acting aggressively, providing a loving but firm model for her to follow, asking for her advice and taking it, and always providing lots of warmth and love.

And if Brazelton doesn't provide enough for you on positive nonpunitive discipline, there is a whole book on it, *Positive Discipline A–Z*, by Jane Nelsen, Lynn Lott, and Stephen Glenn (References, B4). It's not that nurturant parenting ignores discipline. It's just that to a nurturant parent, discipline comes out of nurturance, but it takes a lot of empathy and interaction. It's a lot easier to take a belt and whip a child. But that's a recipe for disaster.

If our bookstores are any guide, nurturant parenting is alive and well and will continue to be. It's a fact that liberals should celebrate. It means that there are plenty of parents and children who have an intuitive understanding of the basis of Nurturant Parent morality and liberal politics.

Childrearing and Politics

Nurturant Parent childrearing practices are superior to Strict Father childrearing practices. But that, in itself, does not show that liberal politics is superior to conservative politics. You might, for example, choose the Nurturant Parent model for your family life and the Strict Father model for politics.

But even with this, there is a problem. The Strict Father model does not work in the primary sphere of childrearing for which it was devised. Its claim to superiority was that it was based on human nature. That claim has been shown to be false.

The metaphorical application of the Strict Father model to

politics is based on the assumption that the Strict Father model works for childrearing, especially in its account of human nature. But since the Strict Father model is wrong about human nature in the childrearing case, there is no reason to think that its assumptions about human nature will be right in the adult case of politics. Indeed, there is every reason to think that its view of human nature will fail as badly for politics as it does for childrearing.

That brings us to the next chapter.

22

The Human Mind

Strict Father morality is not just out of touch with the realities of raising children. It has a problem that goes even deeper. It is out of touch with the realities of the human mind.

To see why, we need to look first at ten deep and necessary assumptions made by Strict Father morality, and then look at what the human mind would have to be like if those ten assumptions could hold. Here are those assumptions:

> *There is a universal, absolute, strict set of rules specifying what is right and what is wrong for all times, all cultures, and all stages of human development.*

If this were not true, there would not be strict moral boundaries, there would be no single straight and narrow path for us all to follow, and there would be no absolute moral standards. This is why conservatives cannot tolerate multiculturalism, which denies this claim, maintaining instead that different cultures may have different rules and standards. Conservatives assume that denying absolute rules and standards is to say that there are no rules and no moral standards at all. The only possibilities they see are moral absolutism or chaos. We will see below that such a dichotomy is false.

Each such rule has a fixed, clear, unequivocal, directly interpretable meaning which does not vary.

If rules have any significant variability of meaning, then moral boundaries and standards are not strict and the "same" rule could legitimately mean different things to different people. If people don't understand "the" rules in the same way, then there is no such thing as "the" rules. There are only different understandings. If the rule is not directly interpretable, then what counts as a moral standard is subject to interpretation, which means it cannot be absolute.

Each moral rule must be literal, and hence must make use of only literal concepts.

If a moral rule is metaphorical, then it is not directly interpretable. In order to know how to follow it, one would have to supply a metaphorical interpretation. But since different metaphorical interpretations are possible, the rule would not be fixed and absolute.

Each human being has access to the fixed, clear, unequivocal meaning of moral rules.

If someone cannot understand exactly what the rule is intended to mean, then punishment for disobedience cannot have the effect of getting the person to follow the rule.

Each rule is general, in that it applies not just to specific people or actions but to whole categories of people and actions.

Rules cannot define general moral standards if they are just about specific individual people and actions. They must be about categories of people and categories of actions.

The categories mentioned in each rule must have fixed definitions and precise boundaries, set for all time and the same in all cultures.

If the definitions of the categories were not absolutely fixed, then the meanings of the rules could vary from person to person, culture to culture, or time to time, and they would no longer be absolute. If the boundaries of the categories were not precise, then the moral standards would not be clear and people would not be able to know exactly what was right and what was wrong.

This is a major point. *Moral absolutism requires conceptual absolutism.* If variability of meaning of any sort is inherent in concepts, then the rules using those concepts are subject to the same variability of meaning. And if that happens, then the whole idea of absolute, universal moral rules becomes impossible.

> *All human beings must be able to understand such rules in order to have the free will to follow them or not.*

You can't make a free choice to do or not to do something if you don't know what that thing is.

> *These rules must be able to be communicated perfectly, from the legitimate authority responsible for enforcement to the person under the obligation to follow them. There must be no variation in meaning between what is said and what is understood.*

People can't obey your orders if they have a different idea than you do of what those orders are.

> *People do things they don't want to do in order to get rewards and avoid punishments. This is just human nature and is part of what it means to be "rational."*

The whole idea of rewarding or punishing people for following or not following rules depends on this being true. If it is not true, then punishing people for breaking rules and offering rewards for following them will have no effect. Without such an effect, authority breaks down.

> *But, for this to be true, people must be able to understand precisely what constitutes a reward and what constitutes a punishment. There must be no meaning variation concerning what rewards and punishments are.*

Here we return to the invariability of meaning. It makes no sense to impose punishments if the meaning of the punishment is not itself clear. If the idea of what the punishment is can vary substantially, then what you think of as a punishment might be understood by someone else as being neutral or even as a reward. Remember Brer Rabbit and the briar patch.

These are minimal conditions on the way human beings must function in order for Strict Father morality to be viable. If these conditions are not *all* met, then that moral system becomes incoherent. For example, suppose that people do not operate generally by reward and punishment. Then the threat of punishment will not be a deterrent, nor the promise of reward an incentive. Without reward and punishment guiding human action, Strict Father morality cannot get off the ground. In short, Strict Father morality requires perfect, precise, literal communication, together with a form of behaviorism.

Thus, Strict Father morality requires that four conditions on the human mind and human behavior must be met:

1. *Absolute categorization:* Everything is either in or out of a category.
2. *Literality:* All moral rules must be literal.
3. *Perfect communication:* The hearer receives exactly the same meaning as the speaker intends to communicate.

4. *Folk behaviorism:* According to human nature, people normally act effectively to get rewards and avoid punishments.

Cognitive science has shown that all of these are false. The human mind simply does not work this way. And it's not that these principles are off just a little. They are all massively false. But, before going on to see why they are false, it is important to see why it is important that they are false.

Categorization

Let us begin with categories. First, categories can be *fuzzy;* they can have shaded borders. What is a rich person? There are clear cases, but no absolute income line clearly demarcates rich from non-rich. There is a gradation. There are no clear boundaries here. One can artificially impose them, of course. But then one could impose them in another way just as well. Consider a moral rule like "The rich should help the poor." If person A does not help person B, it is not always clear whether the rule is being violated.

Fuzzy categories like "rich" and "poor" regularly appear in moral rules. One can always draw lines in one way or another—below this line is poor, above that line is rich. But where one draws these lines is a matter of interpretation and discretion, just what a strictly absolute morality cannot tolerate.

Second, categories can be *radial,* as in the case of a mother. Suppose you have lots of mothers of various kinds. A genetic mother (who donated the egg that formed you). A birth mother (who bore you). Your father's wife at the time of your birth, who raised you. And your father's second wife, your step-mother. How do you know if you have obeyed the commandment "Respect thy mother"? Which mother? All of them? Even the egg donor you've never met?

Even the birth mother you haven't seen since you left the womb? Of course the meaning of mother has changed since the time of the commandment. And that is the point. Meanings change in this way constantly. Most categories are radial. If the concept undergoing change is part of a moral rule, then the rule is not clear and unequivocal. It will require interpretation. But there are always different possibilities for interpretation. And that makes the rule not strict and unequivocal. It means the rule defines not one path but many possible ones.

Third, there are *prototype effects*. Suppose you have a stereotype of athletes as dumb and you are in charge of admissions to a major university. This is, of course, a false stereotype, just as all stereotypes by their very nature are false. Suppose you feel that this places a moral obligation on you not to admit dumb people into the university. Suppose you do, under alumni pressure, admit athletes. Have you violated your self-imposed moral obligation?

The problem is this: Rules contain categories (e.g., dumb people). People usually have stereotypes for thousands of their categories. It is completely normal (though maybe not nice) for people to reason in terms of stereotypes. Because different people have different stereotypes, they will understand a category differently and reason about it differently. That means that they will understand a moral rule containing that category differently. In short, the fact that people really do reason about categories on the basis of stereotypes violates the condition that the meaning of a rule must be invariant from person to person and occasion to occasion. The mind just doesn't work that way.

Incidentally, stereotype-based reasoning is only one form of a much more widespread phenomenon called "prototype-based reasoning." We have seen other examples of prototype-based reasoning in this book. One type is reasoning in terms of ideal cases, as when one thinks about conservatives or liberals in terms of an ideal model of conservatives or liberals. Another type is reasoning on the basis of demons, or anti-ideals. We have seen plenty of cases of demon-based reasoning throughout this book. Another case is called "salient exemplar" reasoning, where one takes a well-known case to stand for a whole category. This is common throughout political and moral discourse.

Fuzzy categories, radial categories, stereotypes, and other forms of prototype-based reasoning all introduce meaning variability. Radial categories are produced, in large measure, because categories do change over time, and their extensions over time are often preserved in radial category structure.

FRAMING

Alternative framing possibilities also provide for forms of everyday variation in meaning. Consider an example from my colleague Charles Fillmore (see References, sec. A3). Suppose you have a friend named Harry who doesn't like to spend much money. You could conceptualize him and describe him in two very different ways. You could say either "He's thrifty" or "He's stingy." Both sentences indicate that he doesn't spend much money, but the first frames that fact in terms of the issue of resource preservation (thrift), while the second frames the issue in terms of generosity (stinginess).

Now imagine an invocation that says: Spend as little money as possible. This is the message that a balanced budget amendment would send to Congress. There are three ways to interpret this invocation: Either "Be thrifty" or "Be stingy" or both. Liberals argue that the government should be thrifty but not stingy. Conservatives argue, on the basis of Strict Father morality, that thriftiness in government is never stinginess, since cutting off government funding just makes people more self-disciplined and self-reliant and so is good for them.

The point is that such an invocation, which is a very real invocation, has two interpretations depending on framing. Moreover, the meaning of that framing depends on worldview, as we have seen throughout this book. But Strict Father morality demands a view of the human mind in which such framing and worldview differences do not and cannot exist. Moral rules, in order to be moral rules, must be understandable in the same way to everybody. The very existence of different worldviews and different modes of framing shows that this is false. The human mind is such that framing differences and worldview differences really do exist, not just here and there in minor ways, but on a truly massive scale. The prohibition "Don't murder babies" may or may not apply to taking a morning-after pill, depending on whether a cluster of cells is framed as a "baby" and taking such a pill is framed as "murder."

Variability in meaning due to framing and worldview differences and to the properties of categories (fuzziness, radial structure, prototypes) creates such a huge meaning variability in normal, everyday human reasoning that the conditions needed for the Strict Father model to be coherent are just not met.

Rewards and Punishments

Such variability also occurs in the understandings of rewards and punishments. Anytime you specify a reward or a punishment, you use human categories that are subject to the same kinds of meaning variability. This means that "rewards" and "punishments" vary in their meaning. Remember the moral of Brer Rabbit: being thrown into the briar patch, which would have been a punishment to others, was a reward to him.

Daniel Kahneman, Amos Tversky, and a small army of co-workers have produced, over two decades, an enormous body of research (see References, sec. A5) detailing how people do not operate by folk behaviorism, that is, according to an objective characterization of what ought to be in their best interests—what ought to count as reward and punishment. Their experiments show in case after case that people just do not reason that way even when it matters a lot to them. Often, the source of that failure is due to the fact that people use other forms of reasoning that get in the way of a reward-punishment form of "rationality"—prototype-based reasoning, alternate framings, worldview differences—which affect how categories of people and events are understood and even affect judgments of simple probability.

The fact is that people do not reason all the time, or even primarily, in terms of maximizing clear and unequivocal rewards and punishments. This fact undermines the principle of the Morality of Reward and Punishment, on which Strict Father morality is based. If punishment isn't always understood as punishment, or if punishment is not usually the basis on which people act, then the whole Strict Father paradigm is undermined. Using punishment to exact obedience to authority and so build self-discipline and self-reliance won't work. And as we saw in the last chapter, it doesn't work in the case of childrearing.

Metaphorical Thought

We have seen throughout this book that people conceptualize a great many things in terms of metaphor, morality itself being one of those things. The fact is that conceptual metaphor exists on a large scale and that it plays an enormous role in moral thought. Take the moral principle that punishment for crimes should be fair. This requires the use of the metaphor of Moral Accounting, and it prompts different accounting schemes around the world. In America, we ask how big a fine or how long a time in what kind of jail is to count as a punishment. The metaphor Well-Being Is Wealth prompts us to try to find a common measure in terms of which we can balance the moral books—balance one kind of harm (assault on well-being) in terms of another. The metaphor of Moral Accounting, being a metaphor, always requires further interpretation if we are to function in terms of it. And the fact that there are many kinds of possible interpretation means that the moral injunction that punishment be fair cannot be followed in just one way. It too has a multiplicity of possible interpretations. Such a multiplicity of interpretations for a moral injunction violates the need of Strict Father morality for a moral rule to have one absolute, universal, true-for-all-times-and-circumstances, clear and unequivocal meaning. The very existence of conceptual metaphor makes Strict Father morality unworkable because it violates the possibility for absolute moral standards.

Imperfect Communication

As for perfect communication, it should be obvious that it simply doesn't work. The failure of perfect communication between liberals and conservatives should show that clearly. The fact of that failure is so prominent in cognitive science and linguistics that it has even become the subject of a best-selling self-help book—Deborah Tannen's *You Just Don't Understand*. Tannen, a former student in my department and now a distinguished professor at Georgetown University, is just one researcher in a field of thousands studying the nature of human discourse and its difficulties. (See References, A4.)

One of the principal results in this discipline is that different people have different principles of indirect speech. Some people are understaters, who say less when they mean more, stop short of the punchline, and let the hearer draw his own conclusion. Others are overstaters, who exaggerate and never miss a punchline or stop short of a conclusion. Different people even have very different views of what constitutes polite conversation. For some people politeness means being indirect, asking a question rather than making a direct request, for example. For others politeness means directness, saying exactly what you mean, no more, no less. And once one gets into the details, the differences in conversational strategies get far more complex than this. Add to this all the meaning variation introduced by framing, worldview differences, metaphor, radial categories, fuzzy categories, and prototype-based reasoning, and you can see why communication is so very far from perfect.

Thus we can see that none of Strict Father morality's requirements for what the human mind must be are actually met by real human minds functioning in real discourses. Strict Father morality is simply out of touch with real minds. Moral absolutism is not true because conceptual absolutism is not true. And moral training by enforcing obedience cannot work because people are not just simple reward-punishment machines.

Relativism

Does the failure of moral absolutism mean total moral relativism? Not at all, no more than the failure of conceptual absolutism means total conceptual relativism. As we saw in our study of the metaphors for morality, those metaphors are not arbitrary or random. They are strongly constrained by what morality is fundamentally about: promoting human well-being. The basic forms of well-being—health, strength, wealth, and so on—constrain the possibilities for metaphors for morality. Even basic forms of parenting-experience—Strict Father and Nurturant Parent—seem to provide a limited range of versions of the overall forms of moral systems. Research in cognitive science on the embodiment of mind shows that, despite enormous possibilities for variation, the variations are not unlimited and not random. They are constrained by various aspects of our biology and our experience functioning in the physical and social world. For an in-depth discussion of why conceptual variation and change does not lead to to anything like total relativism, see References, A2, Lakoff 1987, chap. 18.

Nurturant Parenting and a Nurturant Society

Finally we must take up one more question. Why does the existence of conceptual variation, imperfect communication, and the failure of folk behaviorism not lead to the same problems for Nurturant Parent morality? To see why, return for a moment to childrearing. In the Nurturant Parent model, constant communication, interaction, and discussion are crucial. As Berry Brazelton observes over and over in *Touchpoints,* one must always tell a child why you are doing what you are doing, ask her opinion, ask how she feels, respect her feelings, take her suggestions, while sticking to what you think needs to be done unless your child makes a better suggestion. This process requires constant communication and negotiation of meaning. It assumes that meanings will be different and that communication will be imperfect. It assumes that if you keep communicating, note communication failures, attribute respect and goodwill to both parties, and continue to communicate, you will get to the point where the differences in communication and the variations in meaning won't occur all that much, or matter all that much. What keeps the process of communication going is secure attachment, affection and affectionate behavior, mutual respect, empathy, commitment, clarity of expectations, and trust. That does not apply just to childrearing. It can apply to human interactions in general. That is what overcomes meaning-variations and imperfect communication.

What takes the place of the strict rules of the Strict Father model is clarity of expectations and empathy. What takes the place of reward and punishment is interdependence, communication, and a true desire to remain affectionately connected to those you live with.

FACING DIFFICULTIES

But what happens when the people in your community either want to dominate you or feel no affectionate connections to you or to anyone else? The only answer to date has been to do everything you can to build a nurturant community and extend it more and more to others over time. That is difficult and takes a long time and a lot of commitment and a lot of communication. But the nurturance model in general is difficult to follow and just does take a long time and a lot of commitment and a lot of communication. As with childrearing, there are no easy alternatives. But the Strict Father model is no alternative at all.

Again, as with creating a nurturant family life, it would be unreasonable to expect that creating a nurturant society should be easy or quick. One must be patient and ready to deal with frustration. And one must bear in mind the morality of happiness and self-nurturance. In the midst of frustration, you must find a way to be basically happy and to take care of yourself. If you don't, you will become less nurturant.

Women have known throughout history that nurturance is a way of life. Many men have instinctively learned it from their mothers and their nurturant fathers. But the challenge in contemporary America is to create a nurturant society when a significant portion of that society has been raised either by authoritarian or neglectful parents.

America is between moral worlds and there is only one way to turn.

23

Basic Humanity

Conceptual metaphor is central to moral understanding and moral reasoning. All of Judeo-Christian morality requires the metaphor of Moral Accounting, which is based on understanding Well-Being As Wealth and which allows us to conceptualize such possibilities as paying off moral debts, balancing the moral books, accruing moral credit, and getting one's reward in Heaven.

Without understanding morality as uprightness and evil as a force, we could not conceptualize moral failure as falling, remaining moral as standing up to evil, building moral strength as requiring discipline and self-denial. Without understanding morality as health and immorality as disease, we would not conceptualize immorality as a contagion capable of spreading and requiring an avoidance of contact with immoral people, or demand that it be wiped out. If we did not conceptualize morality as wholeness, we would not reason correspondingly about the crumbling of moral values, the tearing apart of our moral fabric, the decay of traditional morals.

As we have seen throughout this book, a vast amount of

our moral understanding, our moral reasoning, and our moral language comes through such metaphors. But where do those metaphors come from? As we saw in Chapter 3, there is a very simple and straightforward answer. Our metaphors for morality arise from our understanding of experiential well-being: We are better off if we are rich rather than poor, healthy rather than sick, strong rather than weak, whole rather than deteriorating, cared for than not cared for, and so on. States of well-being include health, wealth, strength, wholeness, nurturance, and so on.

The reason that we have the metaphors for morality that we have—both in our culture and in cultures around the world—is that the very notion of morality is founded on experiential well-being and human flourishing. Putting all metaphorical thought aside, what is moral is what promotes experiential well-being in others. Morality is thus correlated with the promotion in others of health, wealth, strength, wholeness, nurturance, and so on. And it is this correlation between morality and aspects of experiential well-being that gives rise to our metaphors for morality. That is, metaphors for morality are grounded in nonmetaphorical experiential morality, in the correlation of morality with promoting strength, wholeness, and health and immorality with promoting weakness, decay, and contagion.

Our abstract system of morality is primarily metaphorical, since it uses metaphors like Moral Accounting, Morality Is Strength, and Morality Is Wholeness. Most of the metaphorical reasoning described in this book makes use of patterns of inference and of language through such metaphors. Because experiential morality is the grounding for all of these metaphors, it is also the grounding for moral understanding and moral reasoning.

The foundation for all abstract, metaphorically conceptualized morality is thus experiential morality; that is, helping, not harming, on the level of direct experience. Without such fundamental forms of experiential well-being—health, wealth, strength, and so on—a metaphorical moral system cannot get started.

Because experiential morality is the basis—the foundation—of all abstract metaphorical moral conceptions, we can ask a very interesting question: Is there ever a conflict between some metaphorical moral system and its basis? Or is there always harmony between a given metaphorical moral system and its basis?

The question is important for a good reason. One sometimes gets the sense that an abstract, metaphorical moral system is somehow losing touch with what morality is all about: losing touch with its basis, experiential morality, losing touch with whether or not human beings are flourishing, losing touch with the health, strength, nurturance, wholeness, and wealth of individual people, on a one-by-one basis.

I believe that we can say something systematic about just what this means and just when it happens. Let us consider the two moral systems we are comparing in this book.

Nurturant Parent morality contains within it something that does not permit it to lose touch with experiential morality. That something is the priority of Moral Empathy, which is the most fundamental metaphor in the entire system of Nurturant Parent morality. Even Moral Nurturance itself presupposes Moral Empathy. The only part of the nurturance system higher than empathy is protectiveness; you don't have empathy with someone trying to kill or harm your child. But otherwise, empathy has the highest priority.

Anyone who puts pure empathy first will empathize with the other fully, put himself in the other's shoes, and therefore not want the other to experience harm—illness, weakness, poverty, misery, deterioration, and so on. It is in this way that Nurturant Parent morality keeps in touch with questions of individual human flourishing, and hence keeps from losing

touch with the foundation of all abstract, metaphorical morality. Even the higher priority given to protection keeps one
in touch with the well-being of those whom you have a
commitment to nurture.

Strict Father morality is rather different in this respect. Moral
Strength, not Empathy, is at the top of its value system.
Right up there are Moral Authority, the Moral Order, and
Retribution (just punishment). Moral Empathy and Nurturance have lower priorities in the Strict Father system. Not
that they are missing. But they must give way to moral authority, moral strength, and retribution.

This means that one does not empathize fully with the
morally weak, such as welfare recipients without the self-
discipline to get a job, unwed mothers without the self-
discipline to refrain from sex, and so on. It means one does
not empathize fully with those who violate moral authority
and break the law; in other words, you don't empathize with
criminals. It also means you don't empathize fully with those
lower in the moral order—species becoming extinct, rainforests, the nonwhite poor of other countries.

The highest metaphors in the system—Moral Strength,
Moral Authority, and Moral Order—therefore do not keep
one in direct touch with human flourishing at the most basic
level of human experience. It matters more in Strict Father
morality that a person is morally weak (lacking in self-
discipline and self-reliance) or violating moral authority (a
criminal) than that he is poor, sick, physically weak, or uncared for.

The Strict Father moral system therefore gives priority to
forms of metaphorical morality—Moral Strength and Moral
Authority—over experiential morality, namely, poverty, illness, physical weakness, and lack of care. This is where this
metaphorical moral system loses touch with the nonmetaphorical, literal, directly experienced foundation of all meta-

phorical moral systems. It is where this system of metaphorical morality loses touch with common humanity.

When an abstract, metaphorical moral system loses touch with the very foundations of any such system, it loses touch with the most basic thing that morality is about.

Overall, Strict Father morality is out of touch. It is out of touch with the realities of raising children. It is out of touch with the nature of the human mind. And it is out of touch with common humanity, with the thing that should be most basic to any moral system.

Strict Father morality is not just unhealthy for children. It is unhealthy for any society. It sets up good vs. evil, us vs. them dichotomies and recommends aggressive punitive action against "them." It divides society into groups that "deserve" reward and punishment, where the grounds on which "they" "deserve" to have pain inflicted on them are essentially subjective and ultimately untenable (as we saw in the last chapter). Strict Father morality thereby breeds a divisive culture of exclusion and blame. It appeals to the worst of human instincts, leading people to stereotype, demonize, and punish the Other—just for being the Other.

Blaming and punishing the Other for being the Other has led, in the worst cases, to the vilest of horrors: the Holocaust and the ghastly tragedies in Bosnia, Rwanda, Somalia, and so many other places. In this country it led to the KKK, and it is what many people fear from the militia movement. But even if there is no killing, a culture of blame is not one that is pleasant or productive to live in. In does not make for a harmonious society or for social progress.

Insofar as Nurturant Parent morality can encourage cooperation and provide the incentive, the training, and the environment in which the largest number of citizens can work together productively and cooperatively, it seems by far the better choice.

Problems for Public Discourse

Public political discourse is so impoverished at present that it cannot accommodate most of what we have been discussing here. It has no adequate moral vocabulary, no adequate analysis of our moral conceptual systems, no way to sensibly discuss the link between the family, morality, and politics—and no way to provide an understanding of why conservatives and liberals have the positions they have.

But the problem with public discourse is even deeper than that. Suppose the central theses of this book are correct, namely:

> Political policies are derived from family-based moralities.
>
> Those family-based moralities are largely constructed from unconscious conceptual metaphors.
>
> Understanding political positions requires understanding how they fit family-based moralities.

Conservative and liberal political positions are impossible to compare on an issue-by-issue basis. Instead, understanding a political position on an issue requires fitting it into an unconscious matrix of family-based morality. The positions are impossible to compare because they presuppose opposite moral systems.

There are no neutral concepts and no neutral language for expressing political positions within a moral context. Conservatives have developed their own partisan moral-political concepts and partisan moral-political language. Liberals have not. The best that can be done for the sake of a balanced discourse is to develop a meta-language—a language about the concepts and language used in morality and politics.

These theses are inconsistent with the very format of news reporting and political discussion in the media. They are also inconsistent with traditional liberal assumptions about political discourse. There are many reasons for this.

First, news reporting assumes that concepts are literal and nonpartisan. But concepts, and the language that expresses them, are typically partisan, especially in the moral and political spheres. The who, what, when, where, and why of news reporting just does not capture the complex partisan differences in metaphorical conceptual structure that lie behind the political positions of conservatives and liberals.

Second, it is assumed that the use of language is neutral, that words are just arbitrary labels for literal ideas. But in morality and politics, that is rarely true. Language is associated with a conceptual system. To use the language of a moral or political conceptual system is to use and to reinforce that conceptual system.

Third, news reporting is issue-oriented, as if political issues could be isolated from the moral matrix in which they are embedded. But political issues are rarely, if ever, isolatable from their moral matrix.

Fourth, the very concept of a traditional debate is at odds with the theses of this book. A debate, by its very nature, combines literalness with issue-orientation. A debate is defined in terms of an isolated issue (like abortion or the balanced budget amendment), which, it is assumed, can be discussed fully and adequately using literal concepts, literal language, and neutral forms of inference. None of this is true. It is also assumed that the terms of the debate are commensurable, that the debaters are in the same conceptual universe. With respect to conservative and liberal politics, this too is false.

Fifth, because language is assumed to be neutral, it is assumed that it is always possible to report a story in neutral terms. But that is not true. To report a story in the language and conceptual system of conservatives is to reinforce and thus give support to the conservative worldview. Where liberals have a language appropriate to their moral politics, the same is true of them. The very choice of discourse form and language to report a story leads to bias. Neutrality is not always possible, though balance may be achievable, at a high cost. Imagine that national news stories were all reported from two opposing moral worldviews. Imagine a box with two columns for each major story, headed "from the conservative worldview" and "from the liberal worldview." Readers might be enlightened, but they might just as easily be confused. And conservatives would cry foul, since, from their worldview, there cannot be another valid and sensible moral worldview.

Sixth, because language is assumed to be neutral, it is taken for granted that the mere use of language cannot put any discussant at a disadvantage. That is also false. Because conservatives have worked out an elaborate language of their moral politics while liberals have not, liberals are put at a disadvantage in any public discourse, and liberals will remain at that disadvantage until they come up with an adequate language to reflect their moral politics.

Seventh, it is assumed by the news media that all viewers, listeners, or readers share the same conceptual system. But that is false. Even the most "objective" reporting is usually done from a particular worldview, one that is typically unconscious and taken for granted by the reporter.

Eighth, the very nature of political discourse in this country makes it difficult to discuss the relationship between morality and politics at all. The separation of church and state has implicitly left the church as the institution that is seen as guarding morality. It has been assumed that all political discussions are issue-oriented and morally neutral. Once one brings morality into issue-oriented discussion, the whole matter of legislating morality is brought to the fore. Conservative support by right-wing churches raises the messy question of how one can discuss morality while maintaining the separation between church and state. Morality is too important to be left to churches. There must be a public discourse on morality, with an adequate vocabulary to show the difference between the moral systems that lie behind liberal and conservative political positions.

Ninth, liberalism itself has a view of discourse that puts it at a disadvantage. Liberalism comes from an Enlightenment tradition of supposedly literal, rational, issue-oriented discourse, a tradition of debate using "neutral" conceptual resources. Most liberals assume that metaphors are just matters of words and rhetoric, or that they cloud the issues, or that metaphors are the stuff of Orwellian language. If liberals are to create an adequate moral discourse to counter conservatives, they must get over their view that all thought is literal and that straightforward rational literal debate on an issue is always possible. That idea is false—empirically false—and if liberals stick to it they will have little hope of constructing a discourse that is a strong moral response to conservative discourse.

In short, public discourse as it currently exists is not very congenial to the discussion of the findings of this study. Analysis of metaphor and the idea of alternative conceptual systems are themselves not part of public discourse. Most people don't even know that they have conceptual systems, much less how they are structured. This does not mean that the characterizations of conservatism and liberalism in this book cannot be discussed publicly. They can and should be. What requires special effort is discussing the unconscious conceptual framework behind the discussion.

—

Afterword, 2002

The Impeachment of Bill Clinton

Events that would have been confusing before I wrote this book now make sense. Let's start with some questions about the Clinton Impeachment.

- Why should the conservatives have seen the Monica Lewinsky affair as an occasion for impeachment? After all, wasn't it a private matter, not an affair of state?
- Why should they have argued as they did? Why did they keep talking about character and teaching children right from wrong?

From the conservative perspective, the President's affair with Monica Lewinsky was the perfect rationale for an attempt at impeachment. It was an affront to Strict Father morality: marital infidelity, abuse of power, betrayal of trust, involvement with an immature younger woman his daughter's age, lies in public, an attempt to cover it up. It was literally a family matter. How could it be made into an impeachable offense on the order of treason or corruption? The

Nation As Family metaphor was made to order. It turned a family matter into an affair of state.

Why did the House Managers look like they did—why did they stand they way they stood, talk as they talked, say what they said? A stricter bunch of Strict Fathers you could hardly find: moralistic, sober, punitive—all trying to look and sound like moral authorities.

Consider the conservative speeches. An excellent example is Wisconsin Republican Representative James Sensenbrenner in his speech to the Judiciary Committee of the House of Representatives.

> Mr. Speaker, most Americans are repelled by the president's actions. The toughest questions I have had to answer have come from parents who agonize over how to explain the president's behavior to their children.
>
> Every parent tries to teach their children the difference between right and wrong, to always tell the truth, and when they make mistakes, to take responsibility and face the consequences of their actions.
>
> President Clinton's actions, every step of the way, have been contrary to those values. But being a bad example is not grounds for impeachment; undermining the rule of law is, frustrating the court's ability to administer justice on private misconduct into an attack of the ability of one of the three branches of our government to impartially administer justice.

And here is an exchange between former Representative Elizabeth Holtzman and Sensenbrenner before the Judiciary Committee:

HOLTZMAN: Mr. Sensenbrenner, I hate to answer a question with a question. But don't you think there's an enormous difference between keeping a dual set of books about bombing of a foreign country without the authorization of Congress and not telling the truth about a private sexual misconduct?

SENSENBRENNER: I think there should be no difference because our perjury and false statements statutes, you know, do not have various levels of perjury. When you do make a false statement, you have to live by the consequences. And I think we all try to teach our kids that one of the things they always should do is always tell the truth.

This was a hearing of the House Judiciary Committee concerning whether the President committed an impeachable offense, on the order of treason or corruption. Why is a conservative member of this committee talking about what you teach your kids? Why does he say that the "hardest questions" of impeachment are about childrearing? Why does he say, "Every parent tries to teach their children the difference between right and wrong, to always tell the truth, and when they make mistakes, to take responsibility and face the consequences of their actions"?

Why does he equate hiding sexual misconduct with such a treasonous act on the world stage as "keeping a dual set of books about bombing of a foreign country without the authorization of Congress"? Why does he say, "When you make a false statement, you have to live by the consequences"? Why does a nationwide TV and radio audience not blink an eyelash when these comments are injected into an impeachment hearing? Why do the media pundits not even find a discussion of child rearing during an impeachment hearing worthy of note?

In the light of this book, the answers are clear. Sensenbrenner is taking for granted the Nation As Family metaphor. He is laying out the Strict Father model of the family, which characterizes the conservative moral worldview. Conservatives are trying to get the public to frame the impeachment hearings from their family-based moral perspective. A metaphor is being set up: members of Congress are the Strict Fathers, the moral authorities. They assume an absolute right and wrong, and they are the moral authorities who determine when that wrong has been committed. Lying is wrong. If your child does wrong, he *must* be punished, because without punishment people would not do what is right and the moral basis of our society would crumble. Punishment is the basis of the "rule of law"—without it, there is no law. Hence, "you must live by the consequences."

In the metaphor, the President is the naughty child. He is obligated to control his sexuality and he hasn't. Control of sexuality is a principal basis of Strict Father family life. The President has shown lack of discipline—he is morally weak, and hence immoral via the metaphor of Moral Strength. Having done wrong, he *must* be punished, and if Congress is to maintain its moral authority, it must insist on punishment. Moral Accounting requires punishment if justice is to be done. In Strict Father morality, the fear of punishment is what makes both children and adults do what is morally right. Without punishment the entire moral system governing society would break down. This is the conservative moral framework as it applies to this case. The conservative congressmen use the "right words" over and over: *rule of law, right from wrong, accept the consequences, administer justice.* The language is meant to evoke the conceptual framework.

President Clinton's strategy was to change the frame. He asked for forgiveness, which makes sense in a Nurturant Parent moral system. He framed it as a family matter, between him and his wife and daughter. In short, he evoked the frame of nurturant family morality.

What is particularly interesting is that over 60 percent of the nation supported Clinton. It was only the hard-core conservatives who favored impeachment. I take this to mean that the "swing voters"—the roughly 20 percent who have access to both the nurturant and strict models—chose the nurturant model to judge this case. This did not go unnoticed by the Bush campaign, which positioned Bush (with considerable success) as a "compassionate conservative"—even though he was a mainline ideological conservative with none of the nurturant moral values at all.

The 2000 Election

"THE MIDDLE"

About 40 percent (plus or minus 2 percent) of the national electorate is consistently strict in its politics and another 40 percent (plus or minus 2 percent) is consistently nurturant in its politics. The portion of the electorate "in the middle" is quite small—about 20 percent (though some estimates go as high as 30 percent). So far as I can tell, this group consists of at least two significant subpopulations:

- "Bi-conceptuals:" People who have both models at their disposal, perhaps used in different parts of their lives, and who may use either model in framing political issues. For example, many blue-collar workers are strict fathers at home, but nurturant in their domestic political priorities. Many executives are strict in their business practices, yet nurturant at home and in their politics.
- "Pragmatists": Pragmatic conservatives and pragmatic progressives. Ideological progressives and conserva-

tives stick firmly to their models, while pragmatic progressives and conservatives are more willing to compromise for practical purposes.

Adding to the complexity, there are different versions of the
strict and nurturant models that emphasize different aspects
of them. Different kinds of progressives and conservatives
have different priorities; there are, for example, fiscal and
social conservatives who place different priorities on different aspects of life. Yet, for all these complexities and variations, the two basic cognitive models dominate American political life. Each provides a relatively simple, and, within its
own terms, consistent, organization for a huge range of political and social issues. Each arises from the central organizing
institution of social life: the family.

The so-called "gender gap" is really a nurturance gap.
White males, on the whole, tend toward strict morality, while
women, on the whole, tend toward nurturant morality. There
is also a "culture gap," with certain states overwhelmingly
strict in their culture (especially in the South and in the
Midwest-to-West region). Other states tend toward a nurturant morality.

What is most important about the concept of "the middle"
is that it is never neutral. Indeed, the very term "middle" is
misleading. It suggests that there is a line from Left to Right
with some people in "the middle" who are neither progressive nor conservative. This is not the case.

First, the idea of a neutral middle misses the pragmatists.
For example, there are pragmatic conservatives who think that
tax cuts are fiscally irresponsible or that a missile shield just
won't work. There are also pragmatic progressives who recognize that parts of the former welfare system weren't working
or who acknowledge that the public school system has serious
problems but don't see vouchers as an answer. These people

are no less conservative or progressive in their ideology than "pure" conservatives and progressives. They just refuse to let ideology override practicality—from their perspective. (Of course, from the purist perspective, they are letting "mere" pragmatic considerations override moral principles.)

Second, consider bi-conceptuals of various kinds. They may apply strict morality in one sphere of life, say finances or foreign policy, but nurturant morality in most domestic issues. Or there are people who have a primary concern with one sphere of life, say race, gender, or sexual orientation. They may be nurturant in that sphere and strict in the others, or the reverse. Again, these people are not neutral in their ideologies; indeed, they may be very fervent in their beliefs.

Third, there are policies that have both strict and nurturant aspects, and people with pure ideologies may focus on those parts that they agree with or those that they disagree with. They may make their minds up in different ways about a given policy without being neutral.

Fourth, there may be policies that can be conceptualized in more than one way by a single individual. For example, education can be seen metaphorically as a business or as a matter of care. A given policy can be seen primarily as either a matter of empathy for children or financial responsibility on the part of the school system. A bi-conceptual who is strict in the financial sphere and nurturant when it comes to children may see such a policy from either perspective, and make one choice or the other or be ambivalent altogether.

It is important to see that people who fall into these complex categories are not "moderate." The term "moderate" suggests a lack of full force behind an ideology and an unwillingness to be overly insistent. None of the above four cases fit this. Appearing moderate is seen as a virtue in much of political life, but is actually rare in practice. There are some true moderates, however: bi-conceptuals who can see both sides of an issue and whose highest concern is avoidance of major conflict (often because they are pragmatists). Nonetheless, I think it is more realistic to replace terms like "center," "middle," and "moderate" with "multi-dimensional" in thinking about the complex 20 percent who are neither pure progressives nor pure conservatives. Much of public discourse is aimed at this multidimensional 20 percent. It is with this population that framing has its greatest effect and must be done with the greatest care.

ELECTION 2000

Gore entered the election with an enormous political advantage; Bush, with an enormous financial advantage. The Bush strategy was to gain and hold the states with a strict morality culture, while largely ignoring the nurturant morality states. His challenge was to get enough of the swing voters from the remaining states to win the electoral vote. Here's how he went about it:

- Bush held on to his conservative base by
  1. Use of body language, voice, and words, all of which told proponents of Strict Father morality from the appropriate states that he was one of them.
  2. Major conservative positions: the tax cut, education standards and vouchers, a foreign policy based on self-interest and maintenance of sovereignty, a punitive approach to crime, and so on.
  3. His choice of Dick Cheney, a trusted conservative, as running mate.

- Bush went after women and other nurturant swing voters with the "compassionate conservative" theme. He co-opted Nurturant Parent language like "Leave no child behind." He was regularly photographed with schoolchildren of mixed race. He spoke of "humility" in foreign policy and he took a directly moral tone— "returning integrity to the White House."

Those who knew Bush and Cheney's histories knew they were strict conservatives—true believers. But many voters were fooled by the nurturant façade.

It was Gore's election to lose and he did everything he could to lose it.

- While Bush held on to his extreme conservative base, Gore did nothing to hold on to the most progressive of Democrats, letting Green Party candidate Ralph Nader make fatal inroads into that base.
- Instead of attaching himself to the enormously popular Clinton, he distanced himself from Clinton and would not even let Clinton campaign for him in crucial states.
- Where Clinton projected empathy in public, Gore was wooden, and appeared distant and self-righteous. Despite constant comment in the press, Gore did nothing to change his gesturing system.
- Where both Clinton and Bush understood the symbolic dimensions of politics, Gore missed it almost entirely and depended on policy prescriptions to carry the day.
- While Bush courted women and other nurturant swing voters with "compassionate conservatism" and other slogans, Gore ignored the nurturant vote and campaigned instead on the theme "I'll fight for you." This was a disastrous choice. It not only ignored the nurturance theme, it contradicted that theme with the word "fight." It also suggested powerlessness; it went against the very idea of the president as powerful, suggesting that being president wouldn't be enough, that he would still be struggling. It actually made Gore sound like a wimp, not the warrior he was trying to portray himself as.

- Moreover, Gore did not understand what strategy Bush was using. He did not know enough to attack Bush on the compassion issue, to portray himself as the true candidate of compassion.
- Where Bush campaigned on moral issues, Gore did not and thus lost the moral high ground. The charge that he was dishonest and a manipulating politician who just wanted to win seemed to stick in the absence of a clearly delineated moral standpoint.
- Gore misunderstood the nature of the televised debates. He assumed that debates were won by scoring policy points, which should have made him the superior debater. Bush won by framing issues to his advantage, while Gore did not know how to reframe an issue. Gore responded instead with facts, figures, and policies, which let Bush confound him by retaliating with his own facts, figures, and policies—and the public, of course, could not tell the difference. All issues of statistics and math became "fuzzy math."
- Gore attacked the Bush tax plan by pointing out that most of the savings would go to the "top one percent" of taxpayers. The assumption here was that all lower- and middle-class voters (a) would be voting their self-interest and (b) would be resentful of the rich getting richer. Mistake (a) was that lower- and middle-class conservatives, following Strict Father morality, believe that the rich deserve what they get, that they should keep what they earn, and that the tax cut would be moral. They were voting morality, not self-interest (see Chapter 10, pp. 189–92). Mistake (b) was that a great many lower- and middle-class voters admire and want to emulate the rich, and a high percentage (over 25 percent) actually thought they were, or eventually would be, in the top one percent. The top one percent ar-

gument is always pointless. It works with ideological liberals, but Gore already had their votes. Overall, it fails.

In short, there was a major cognitive component to Gore's loss of what should have been a sure thing. Gore and those running his campaign did not understand the content of this book—the cognitive and symbolic dimension of politics—while Bush and his campaign managers did.

THE FLORIDA VOTE COUNT

Many people think that words are "mere words"—just labels for aspects of an objective reality. Cognitive linguists know better. The way we use language both reflects how we think about something, and can impose on other people a way of seeing the world. Language matters, and the ideas (often metaphorical ideas) expressed by language matter.

Here are five ways of conceptualizing and talking about an election.

1. A horse race. This is the most common metaphor for an election used in the media. The metaphor carries over certain common inferences about horse races to elections. Thus, we expect that:

    i. Everyone starts off from the same place and has to run the same course.
    ii. There are rules and people will obey them.
    iii. The more able candidate wins the election, just as the faster horse wins the race.
    iv. There are neutral officials who guarantee that the race is fair and who judge the winner in the case of a close race.

v. There is always a winner and the winner of the race is the candidate that is "ahead when the horses cross the finish line."

vi. It doesn't matter how you win or much you win by as long as you win.

vii. There is also an etiquette of racing. The loser is supposed to take his loss graciously and not contest the call of the neutral judges. If he does not, he is a "sore loser."

2. A football game.

i. The game is rough. It is supposed to be rough. Players get hurt. Toughness is required to win. You are supposed to play as rough as you can in order to win. A certain amount of dirty playing is expected.

ii. There is a quarterback calling the plays. The other players have to do what he says, or they are taken out of the game or thrown off the team.

iii. The game ends after a fixed time. The team ahead when the game ends wins.

iv. It is a time-honored, respectable strategy for the team that is ahead toward the end of the game to "run out the clock," that is, to use up as much time as possible to keep the other team from scoring.

v. Again, there are neutral judges who guarantee that the play of the game is fair and who determine the winner of a close game.

vi. Again, it doesn't matter how you win or how much you win by as long as you win.

vii. And again, the etiquette required is that the

loser be gracious in defeat and not be a "sore loser."

3. A war.

i. This is a fight for survival, and for who gets to rule.
ii. Both sides are "out for blood."
iii. The candidate with the greater firepower, the best strategy, and command of the terrain will win—not necessarily the one with the most ability.
iv. Fine points of ethics may be reasonably overlooked when survival is at stake.
v. Again, it doesn't matter how you win or how much you win by as long as you win.

4. A legal process for democratically establishing "the will of the people."

i. Maximum voter participation is essential to establish "the will of the people." Anything interfering with voter registration or actual voting delegitimates the election, should be considered illegal, and should be legally contestable.
ii. "The intent of the voter" is foremost in establishing "the will of the people." Anything interfering with voter intent being registered in such a way as to affect the outcome of the election should be illegal and make the election contested.
iii. Every vote should be counted.
iv. The vote count should be carried out impartially.
v. The election isn't over until the legal process has run its course.
vi. If the election does not or cannot establish the will of the people, a new election should be held.

5. A political process that is just another aspect of a tough, competitive world.

   i. Everyone naturally seeks their self-interest.
   ii. The individual is responsible for making his own way.
   iii. Those disciplined and powerful enough to win competitions like elections rule—and should rule.

All of these ways of conceptualizing an election entered into the Florida vote count in the 2000 election. The media used the horse race metaphor as it always does. There are actually two versions of the metaphor. In the overall race metaphor, the race extends from the time candidates "start to run" to the actual election, and who is "ahead" is determined by polls. In the vote count version of the race metaphor, the race begins when the results start coming in, and who is "ahead" is determined first by exit polls and finally by "official vote counts" on election night (when no recounts have yet been done). A race that is "too close to call" on election night is something like a photo finish where impartial judges later look carefully at the photos made at the finish line and decide the winner.

In using the horse race metaphor for the election, the media gave a major advantage to Bush. It imposed an understanding of the election based on the metaphorical principle: *There is always a winner and the winner of the race is the candidate that is "ahead when the horses cross the finish line"* (Metaphor 1, line v above). Moreover, each time a partial recount was made, the vote-count version of the horse race metaphor was used: each recount, however partial, was another race. Bush's people kept repeating this metaphorical inference over and over: Bush "won on election night" and he "won on every recount." When Gore's people protested that the election was a legal process that wasn't over and that no one had won anything until the legal process was completed, we entered into a national "metaphor war." Conservatives, sticking with the race metaphor, called Gore a "sore loser" for protesting.

Interestingly, the choice of election metaphors not only fit each side's partisan interests, but they also fit the moral worldviews of liberals and conservatives. In Strict Father morality, competition is not a necessary evil, but a positive good: it is an absolutely necessary condition for morality to exist. Without competition, there would be no need for discipline, and without discipline, people would not follow moral laws (see Chapter 5, p. 69). For this reason, Metaphors 1, 2, 3, and 5—horse race, football, war, and tough-world politics— are all conservative metaphors. Thus, the conservative delaying tactics in counting the votes and in getting legal decisions were seen as legitimate by conservatives, a form of "running out the clock" (football metaphor) when they were "ahead."

Liberals, in accordance with Nurturant Parent morality, naturally chose Metaphor 4—the legal process for democratically determining the will of the people. This follows from the nurturant notion of fairness (see Chapter 6, pp. 123–24), and from empathy with anyone denied a political voice. It is no accident that the liberal Florida Supreme Court should have used "voter intent" as the basis for a recount.

It should also be said that the war metaphor is used commonly by political professionals in both camps. Both Democrats and Republicans have their "war rooms."

Metaphor 5—the tough world metaphor—played an especially important role in the election. The conservative reaction to the confusion over the butterfly ballot in West Palm Beach was that if you can't figure out the ballot, your vote shouldn't count. Their reaction to the hanging chads was that, if you can't punch a clean hole in your ballot and check it, then your vote shouldn't count. In other words, if you don't have enough discipline, too bad for you. You have to accept the consequences. Even Justice Sandra Day O'Connor was reported to have made such a remark at a private gathering, and in oral argument before the court, she remarked, "Well, why isn't the standard the one that voters are instructed to follow, for goodness sakes? I mean, it couldn't be clearer." The thousands of blacks who were either purged from the voting rolls or falsely told the polls had closed were seen by many conservatives as not having been disciplined enough to make sure they voted.

Metaphor 5 was also the basis for the feeling of moral superiority on the part of conservatives when Katherine Harris and other conservative election officials used their power openly to Bush's advantage. Because Republicans controlled the offices of the governor and secretary of state, they controlled the everyday legal mechanism governing the election. What they did *was* the law, and hence Bush's people could say that they supported the "rule of law." The term "rule of law" is important in Strict Father morality, since "law" defines what is right and wrong and "rule of law" characterizes moral authority.

Finally, the Supreme Court's 5–4 ruling by conservative justices that made Bush president was an instance of Metaphor 5. The liberal metaphor—Metaphor 4—depends on the legal process, which, in the metaphor, is an idealized legal process. But in real life, the law of the land is decided by judges, who are political appointees, through the use of naked power. The conservative justices had no qualms about using their power first to stop the vote count with Bush "ahead" and then, since the count could not be completed on time, issued an arbitrary ruling that was tantamount to simply choosing Bush as president. Their ruling came right out of the Strict Father moral system:

> Florida's "intent of the voter" standard for judging ballots is unconstitutional under the equal protection clause because of "the absence of specific *standards* to ensure its equal application." It further held that the "formulation of uniform *rules* to determine intent based on these recurring circumstances is practicable and, we conclude, necessary."

"Standards" and "rules" are at the heart of Strict Father morality: rules distinguish right from wrong and standards define what counts as being disciplined enough.

The arbitrariness of the ruling is indicated by the fact that it was explicitly made not to set a precedent, but to apply only to this one case. The court had to limit the ruling to this case, and without such a limit, the ruling could have thrown out a huge number of everyday laws in the country. The reason is that "intent" is decided in a myriad of cases, from negligence to discrimination, without any "uniform rules." Intent is a "standard" determined every day in courts all over the country without uniform rules. Indeed, the Florida electoral system itself did not have "uniform rules" since it allowed different kinds of voting machines, whose job is to determine voter intent, with the more accurate ones chosen not on the basis of uniform rules, but on the basis of the wealth of the particular county (since the wealthiest counties had the best voting machines, namely, optical scanners). If their ruling had been general, as their rulings typically are, it would have thrown out the Florida election and made Gore president.

The conservative Supreme Court justices acted on the basis of Metaphor 5—the tough world metaphor. They had the power and they used it, arbitrarily, to stop a vote count and to make one of their own president. Did they feel bad about it? Did they think they had done something immoral? Certainly not! First, what they did was legal. Indeed, they have the power to determine what is to count as "legal." Then, the Strict Father moral system makes promoting Strict Father morality itself the highest form of moral action (see Chapter 9, p. 166). Metaphor 5, which comes right out of Strict Father morality, allowed them to use their power to stop the vote count before Gore could gain enough votes to win. And they acted in accord with the media's Major Election Metaphor— the horse race—which ruled out any significant media criticism since Bush was "ahead" according to this metaphor on election night when the horse race was over.

Liberals attacked the ruling by the five conservative justices as contradicting their professed views on states' rights. If the justices were consistent, it was argued, the court would have invoked states' rights and upheld the jurisdiction of the Florida Supreme Court. This is a common liberal mistake: taking conservatives to task for supposedly contradicting themselves, say, on states' rights and interference by the federal government. The reason why this is a mistake is a straightforward matter of moral politics. The overriding concern of conservatives is Strict Father morality; morality is always the number one concern. Particular political issues are understood within this moral framework and defined relative to it. Conservatives are not against big government and in favor of states' rights in any absolute sense, but only relative to their moral system. There is no contradiction here at all from a conservative perspective. Conservatives are not against big government when use of the power of the federal government fits their moral system. In this case, it did.

If Gore had won, the narrowness of the victory would, under Metaphor 4, have indicated that he did not have the will of the people solidly behind him. As a good liberal, he would have sought some sort of accommodation with conservatives. And many liberals thought that Bush, with so narrow a victory, should have sought accommodation with liberals, since *he* did not have the will of the people behind him. They were shocked by Bush's lack of real bipartisanship. They shouldn't have been. As a conservative, his election metaphors required no such thing. According to Bush's metaphors, it doesn't matter how you win or how much you win by as long as you win. He set forth a thoroughly conservative agenda as though he had a major mandate, just as one would have expected from an examination of the Strict Father moral system.

The First Months of the Bush Administration

Strict Father morality describes the Bush administration's actions to a T. The Bush team's understanding of the election meant that they would take victory, no matter how narrow or what kind, as a mandate, not an occasion for bipartisanship. Moreover, as *moral* conservatives, real bipartisanship would have meant compromising their moral views.

What is important to understand about the Bush administration, at least in its first few months, is that its major actions were not arbitrarily chosen. They are general, higher-order policies that affect thousands of particular issues. The conservative strategy is that winning a single-higher-order victory provides victories in thousands of special cases. An example would be controlling judicial appointments. If you appoint conservative judges, they will make thousands of decisions in the conservative direction. Thus Bush's executive order ending review of federal judicial appointments by the American Bar Association means that there will be less information and justification available for liberals to stop the appointment of ideologically biased conservative judges. Bush's appointments of John Ashcroft as attorney general and Ted Olson as solicitor general means that many more conservative judges will be proposed. Given the limit on the number of vetoes by individual senators that will actually occur on the Senate Judiciary Committee, this means the courts will move in a conservative direction.

The 1.3 trillion-dollar tax cut is another higher-order strategic move in support of Strict Father morality. If the cuts hold up over the years, there will simply be little money available for social programs, which are immoral in Strict Father morality (see Chapter 10). Instead of having to work to eliminate them one by one, the tax cut eliminates lots of them all at once. It also rewards what Dan Quayle referred to as "the best people" (Chapter 10, p. 189), according to Strict Father morality, namely, those who have used their discipline, engaged in competition, sought their self-interest, and become self-reliant (read "wealthy"). It is a single overall reward, not a bunch of separate small tax breaks. The tax cut is thus doubly moral from the perspective of the conservative moral system.

ENERGY AND THE ENVIRONMENT

The Bush energy plan was part of another such higher-order strategy. The strategy was to frame energy as the heart of the economy while destroying environmentalism in the process. Here is how the strategy was carried out in the first months of the administration.

- Put pro-business, pro-energy-development people in charge of the most environmentally sensitive agencies: the Interior Department (Gale Norton) and the EPA (Christie Whitman).

- Cut funds for research and development on conservation (e.g., fuel economy, which would vastly lessen the need for oil) and environmentally responsible energy sources (biomass, wind, solar, and so on).
- Announce a national energy supply crisis and call it a matter of national security. Develop a plan to respond to the "crisis."
- Frame the "crisis" so that environmentalists are defined as the problem: their regulations impede the development of supply.
- Appoint commissioners to the Federal Energy Regulatory Commission (FERC) who would refuse to cap electricity prices overall, even though FERC's mission is to guarantee reasonable energy prices.

The Bush administration took advantage of the market manipulation by energy companies in California, announcing a national "energy supply crisis" despite the fact that peak energy use in California had barely risen, if at all, according to Christian Berthelsen and Scott Winokur in the article "Soaring Electric Use More Fiction than Fact," published in the *San Francisco Chronicle* 11 March 2001. Energy companies were simply cutting off supply to drive up prices.

Were Bush and Cheney just acting to make their friends and political supporters rich, as many liberals alleged? There is a case to be made for that, but there is much more to it.

THE CONSERVATIVE WAY OF LIFE

From the perspective of conservative morality, nature exists for human exploitation (see Chapter 12). Environmental regulations get in the way of profits and the use of private property, rewards for "the best people"—those who are disciplined, who pursue their moral self-interest in the marketplace, and who are able to succeed. The Energy Crisis Frame establishes environmentalism as the Problem, the Bad Guy standing in the way. If this frame can be accepted generally, then environmentalism can be defeated in general, not one issue at a time.

Liberals made jokes when Bush and Cheney came out against conservation as public policy. *What? Conservatives who don't conserve!* But look at conservation from the perspective of Strict Father morality. Conservation does not provide rewards for "the best people." It is not an incentive to be disciplined and industrious. Such incentives are profits and consumer goods. Environmentalism (seeing nature as having overriding inherent value) and conservation are values that fly in the face of this form of Strict Father morality. Cheney has referred to conservation as a "private virtue," casting conservation as asceticism and environmentalists as ascetics who would rob Americans of their hard-earned consumer goods.

Bush and Cheney have framed energy as being at the heart of a consumerist society, a society organized according to Strict Father values. Their proposal to build 1,300 new power plants—an average of one a week for 20 years—is a proposal to place those values at the center of American life.

Bush, during his first few months in office, announced that he saw the federal government as serving the needs of business. "Business" does not mean the people working for businesses; they are only "human resources." It means the entrepreneurs, the owners, the investors, and managers of businesses—the people who are disciplined, seek their self-interest in the competitive marketplace, and succeed; in other words, the "best people." This was an acknowledgment that Bush is not against "big government" as long as it serves Strict Father morality. On energy, he is making heavy use of the apparatus of government to benefit business: FERC, the Energy Department, the Interior Department, the EPA, and so on. No conservatives are accusing him of "big government," even though he is using it, because he is using it for conservative moral ends—to impose on the country a conservative way of life.

From this higher-level perspective, many specific early actions of the Bush administration fall into place.

- Arsenic in the drinking water: Bush acted to undo Clinton's executive order for stricter controls on arsenic in drinking water. High levels of arsenic contribute to cancer and leukemia, and are especially dangerous to children. Arsenic is also a byproduct of mining, and mining interests can save a lot of money by not having to filter out the arsenic produced by their operations.
- The administration acted to free mine owners from having to post bonds to ensure that they will clean up their messes. Again, this saves them money, though at the cost of possible environmental devastation if they do not clean up.
- Clinton issued an executive order banning roads in a huge range of our national forests. Roads interfere with wildlife habitat, while providing access for development, which could ruin the forests. Bush is seeking to overturn the Clinton order in as many places as possible, so that oil and mining interests can exploit these areas.
- Bush is seeking to allow oil drilling in the Alaskan National Wildlife Refuge, the most pristine natural area left in America. Allowing drilling would, environmentalists argue, destroy the refuge. Compared to Alaska's North Slope, there isn't all that much oil in the Refuge.

But the Wildlife Refuge has become a symbolic issue. If the conservatives can drill there, then they should be able to drill anywhere.

- The conservative congress passed and Bush signed a bill overturning existing ergonomic standards for workplace safety as too expensive for industry.
- The Bush energy plan supports both nuclear power and coal as energy sources. Coal pollutes, producing a vast amount of carbon dioxide and other pollutants. Among these is mercury, which gets into waterways and concentrates in fish, making many kinds of fish into dangerous carcinogens. Nuclear power plants produce nuclear waste, which is one of the most toxic substances on earth and remains deadly for over 10,000 years. No safe place to store nuclear waste has yet been found. Moreover, the proposed site at Yucca Mountain is in an earthquake zone, above underground streams, with potential danger of very long-term contamination of underground water for hundreds of miles. Besides, nuclear energy requires enormous federal subsidies. But both coal and nuclear energy provide avenues for private profit—rewards for enterprising businessmen.

In all of these cases, the Bush administration acted consistently to promote a conservative way of life, centered around Strict Father morality.

FOREIGN POLICY

Clinton's foreign policy had two pillars:

1. Economic Globalization and Interdependence. The theory here was that (1) countries that are economically dependent on each other do not go to war; (2) democracies do not go to war; and (3) countries with free mar-

kets that engage in free trade are more likely to be, or eventually to become, democratic. Part of the idea is to eliminate conflict with former or current adversaries (e.g., Russia, China, North Korea) by bringing them into trade relations and trade dependencies. Moreover, since the U.S. is wealthy and has lots of economic know-how, it is at an advantage in global trade. In short, you can promote world peace and prosperity while doing well economically.

2. International Moral Norms (of a limited kind). When things work well in international relations, it is because certain moral norms are adhered to. Maintaining such moral norms is thus crucial to a smoothly functioning world system. Violations of such moral norms are not to be tolerated. Such violations include genocide and ethnic cleansing, state terror against a state's own people, and starvation as a political strategy. Clinton adapted the military to help deal with such situations— e.g., in Bosnia, Kosovo, and Haiti—with the goal of helping to establish viable democratic governments.

A conservative approach to foreign policy based on Strict Father morality is very different, and we see it in the Bush administration. From Moral Self-Interest (Chapter 5, pp. 94– 96), there follows a policy of American self-interest in international affairs. From the Moral Order (Chapter 5, pp. 81– 84), we get the idea of America as being morally superior to other nations and hence deserving greater authority. From this it follows that America should not give up any of its sovereignty. And because of the central role of profit-seeking in Strict Father morality, governments that are still communist (e.g., China and North Korea) must be seen as fundamentally immoral adversaries. Finally, there is the idea of spreading the conservative way of life from America to other countries.

The following actions of the Bush administration in its early months are examples of such a foreign policy.

- One of the first actions of the administration was the banning of birth control information in federally funded clinics abroad. Here we have the spreading of the conservative way of life to other countries. It has the effect of limiting information on abortion, of course, but it also serves to make it harder for women to take control of their lives.

- The administration, in defining China and North Korea as adversaries rather than as possible partners, is thus reincarnating the old adversarial relationship with communist countries. For North Korea, the result is that (a) there will not be a treaty in which North Korea gives up making missiles and that (b) an accommodation between North and South Korea will be unlikely. In expanding military support to Taiwan, the U.S. is acting as an adversary to China.

  Actually, there is a split in the administration as to whether self-interest (expanding trade) should override anti-communist policies in the case of China. Both are conservative values, but they can come into conflict where China is concerned.

- Having created adversaries who have missiles, the Bush-Cheney administration now seeks an anti-missile defense system. The arguments against it, of course, fall on deaf ears, (a) that it will start a new arms race, (b) that the real threat is from terrorism, not missiles, and (c) that it just won't work.

- In order to have a new anti-missile defense system, the U.S. had to get out of the anti-ballistic missile (ABM) treaty.

- Declaring that carbon dioxide is not a pollutant was necessary for the administration's domestic energy plan. It took the U.S. out of the Kyoto accords. But the Bush administration was against them anyway on grounds of self-interest, saying that accepting the accords and acting on them would hurt the American economy "too much." Other studies had indicated that the economic harm would not be great.

- As an extension of the conservative way of life to other countries, the administration negotiated the building of power plants in Mexico and Canada to supply power to the U.S. They will produce pollution in those countries. They may well be built through American investment with profits going to American energy companies.

During the first few months of the Bush administration, Democrats had very limited success in standing up to the conservative juggernaut. They did an especially bad job of explaining to the public what the Bush administration has been up to, and an even worse job of putting forth an integrated, high-level, well-thought-out, and articulated program of their own. Furthermore, Democratic leaders have shown no sign that they understand the role of moral politics in the conservative agenda. The defection of Senator James Jeffords of Vermont to the Democratic ranks shifted the balance of power in the Senate. Only this has slowed down the conservative Republicans.

The Challenge for Progressives

THE THINK TANK GAP

The Bush administration came into power well-organized, with sophisticated high-level plans for promoting conservative values in every sphere of life. That was no accident. Conservative think tanks had been planning the details of a massive conservative takeover of the federal government and a corresponding cultural shift for years. Conservative intellectuals associated with these think tanks have done their job very effectively. They have worked out, implicitly if not explicitly, the relationship between the family, morality, religion, and politics within the conservative worldview. They have thought through how issues are related to one another and how higher-level framings of issues can greatly increase their political effect.

Over the past thirty years, conservatives have spent a huge amount of money—well over a billion dollars in the 1990s alone—to support their intellectuals in such think tanks. Conservative intellectual leaders are linked up with policymakers—leaders in government, business, religion, conservative causes, and the media. Career development for conservative intellectuals is well-funded. Their research is given financial support and is widely distributed to the media through a vast network.[1]

Because of the way conservative think tanks are funded—through large general block grants and virtually guaranteed long-term funding—conservative intellectuals can work on long-term, high-level strategies that cover the whole spectrum of issues.

Liberal think tanks and other organizations are not only outfunded four-to-one, they are also organized in a self-defeating manner. There are three general types: advocacy, policy, and monitoring the other side. The advocacy and policy organizations generally work issue-by-issue. Few are engaged in long-term, high-level thinking, partly because of the issue-by-issue orientation, partly because they are kept busy responding to the current week's conservative assaults, and partly because they constantly have to pursue funding.

The funding priorities of liberal foundations and other funders are also self-defeating. They tend to be program-oriented (issue-by-issue) and relatively short-term with no guarantee of refunding. Moreover, they tend not to give money for career development or infrastructure. And liberal organizations tend *not to support their intellectuals!* In short, they are doing just the opposite of what they should be doing if they are to counter the conservatives' successes.

If this were a rational issue, just pointing out the problem and its horrendous consequences would be enough to get liberal-oriented foundations and funders to change their way of doing business. But it is not a rational issue—it is a matter of moral politics. The roots of the problem are deep and not at all obvious. They lie within nurturant morality itself. The liberal moral system itself is working against the possibilities of its own success.

The reason can be found in Chapter 9, pp. 163–67, in the discussion of the highest priorities of the two moral systems.

- Strict Father morality has as its highest priorities:
  i. Promoting Strict Father morality in general.
  ii. Promoting self-discipline, individual responsibility, and self-reliance.

- Nurturant Parent morality has as its highest priorities:
  i. Empathetic behavior and promoting fairness.
  ii. Helping those who cannot help themselves.

Because conservatism's highest moral priority is to promote the moral system itself, conservatives find it natural to support ideas, infrastructure, and career development. Because its other highest priority is to promote self-discipline, individual responsibility, and self-reliance, conservatives find it natural to support American enterprise, competitiveness, the free market system, and so on.

Because liberalism's highest moral priority is empathetic behavior and promoting fairness, it is natural for liberals to support advocacy for the downtrodden and oppressed. Because its other highest moral priority is to help people who can't help themselves, it is natural for liberal funders to insist that their funding go as close to the needy as possible, to the grassroots, not to infrastructure or career development—*and certainly not to intellectuals!* The result is that conservative funders have been promoting conservatism, while liberal funders have not been promoting liberalism.

Though conservative think tanks get a lot of money, their money does not come from the wealthiest foundations by any means. There are plenty of liberals with enough money to match the conservatives. Wealthy liberals, however, want their money to go as directly as possible to the downtrodden and oppressed, with nothing significant designated for infrastructure, career development, or their intellectuals. From a position external to the liberal moral system, this seems irrational and self-defeating. But from *inside* the moral system it seems natural.

WHAT LIBERALS NEED TO DO

There is Black Pride and Gay Pride, but no Liberal Pride or Progressive Pride. It's about time liberals got some.

Liberals have a moral system. It is described explicitly in this book. It is organized not around adherence to specific rules, but around a higher principle: Help, don't harm! It is an ethics of care, centering around empathy together with responsibility, both for oneself and others. It is the moral basis of all the specific programs and issues. It is the reason that liberals find virtually all of the moves of the Bush administration abhorrent. Liberals feel moral outrage, but cannot express it because they shy away from the very idea of morality.

Conservatives have taken the term "moral" for themselves and liberals have let them keep it! It is time to take it back. "Morality" is a powerful idea. Our greatest leaders have been moral leaders. The great issues are not policy issues but *moral* issues! Wonderful words and expressions like *Freedom, Liberty, Integrity, the Rule of Law*, and *the American Way of Life* have come to have a conservative connotation. Right now conservatives own these words and it is time to take them back, to give them proper meanings again within nurturant morality.

Liberals have been fooled into arguing on conservative grounds, fooled into defending what conservatives argue against, e.g., "big government" and "government spending." Those aren't really major issues in themselves, but only when seen through a conservative moral lens. As we have seen, conservatives are all too happy to use big government and government spending on what they perceive to be a moral cause. These terms cannot be taken literally as if they were free-floating; instead, they have come to be defined by conservatives relative to a Strict Father worldview.

This brings up some lessons to be learned from cognitive linguistics.

- Words are defined relative to conceptual frames. Words evoke frames, and if you want to evoke the right frames, you need the right words.
- To use the other side's words is to accept their framing of the issues.
- Higher-level moral frames limit the scope of the frames defining particular issues.

- To negate a frame is to accept that frame. Example: To carry out the instruction "Don't think of an elephant" you have to think of an elephant.
- Rebuttal is not reframing. You have to impose your own framing before you can successfully rebut.
- The facts themselves won't set you free. You have to frame facts properly before they can have the meaning you want them to convey.

These are some of the principles of strategic frame analysis. Given the conservative think tank advantage, liberals need to master the art of strategic framing—and reframing—as soon as possible.

Here are two examples of the kind of reframing of public discourse that is needed.

THE FIRST REFRAMING: THE TWO-TIER ECONOMY

There is a persistent and terribly damaging myth about our economy, namely, that in the American economy poverty can, in principle, be eliminated—if only there is better education, more jobs, more opportunity, and if people will just work hard, save, invest, and pull themselves up by their bootstraps. This is simply false. Our economy as it is presently structured *requires* substantial poverty.

The present American economy *requires* that certain jobs have low wages: cleaning houses, caring for children, preparing fast food, picking vegetables, waiting on tables, doing heavy labor, washing dishes, washing cars, gardening, checking groceries, and so on. In order to support the lifestyles of three-quarters of our population, one-quarter of our workforce must be paid low wages. These are the people who make two-income families possible, because they take care of the house and the children, allow fast-food outlets, restaurants, and hotels to exist, and perform other tedious, unpleasant, unsafe, and physically difficult jobs that support middle-, upper-middle, and upper-class life.

It is a myth that *all* the people so employed can lift themselves up by their bootstraps, get educated, spend thriftily, save, invest, and get out of poverty—that is, to get decent housing in a safe neighborhood, adequate food, health care, and education for their children. Even if *all* the present lower-tier workers moved into the upper tier, the country would still need a quarter of the population, working at low wages, to take care of the children, clean the house, work in fast-food places, pick the lettuce, weed the lawns, wait on tables, wash the cars, and so on. This economy absolutely relies on hard-working people whose pay does not reflect their contribution to the economy.

In short, those on the ground floor of our economy are holding up those on the upper floors—and they work hard to do so. But the structure of our economy does not allow their pay to be commensurate with their contribution to the economy as a whole.

A free-market economy is one in which labor is seen as a commodity that people should be able to sell for what it is worth. But in our economy, individual employers cannot, for the most part, afford to pay lower-tier workers a wage that reflects what they contribute to the economy overall.

In an important sense, lower-tier workers are working for the economy as a whole, since they make upper-tier lifestyles and incomes possible. In a well-run market, people should be able to get what their labor is worth. But we do not have a well-run market. What is needed is a market correction— a way that the economy as a whole can reward those whose labor it depends on but *cannot* adequately pay. The mechanism is simple: a negative income tax (that is, a serious expansion of the earned income tax credit).

What do lower-tier workers deserve for making middle- and upper-class lifestyles possible? What is the *least* they deserve? *Adequate health care, adequate nutrition, decent housing, and full access to education.* Can the economy as a whole afford it? I suspect so, but the question has not been asked, at least not properly: Can we afford a moral economy—a fair, well-run economy in which people are paid what they have earned, that is, what their work is worth to the economy as a whole? Can we at least provide a "moral minimum"— the *least* that lower-tier workers deserve? Anything less is simply immoral, and a market that pays less when it could do much better is not a well-run market. This is a national discussion we need to have. It is a discussion that makes clear that markets are not forces of nature; they do not just happen; they are not totally "free"; they are constructed and run, and the question we must ask is *how* they should be run.

Reframing 2: The Ecology of Energy and the Real American Way of Life

The Bush administration in its first months framed environmentalism as the Problem in the Energy Crisis. This frame needs to be turned upside down. Energy is a matter of ecology. There is a whole landscape of energy sources, clean and safe ones—biomass, wind, solar, the tides—with adequate technology already available or in sight waiting to be developed. Energy is part of a *system*, an ecological system. Energy is not an isolated issue. It is a *Health* issue: the purity of the air we breathe, the water we drink, and the food we eat are all Health issues. How air smells and water tastes are *Quality of Life* issues, as is the role of nature and beauty in our lives. Energy is also a *Moral* issue. How we get our energy and whether our government and businesses allow for its efficient use so that we minimally despoil the earth are questions of morality. We have a responsibility to take care of the Earth, a religious responsibility if you believe in God and an ethical one if you do not. Energy is also a *Scientific* issue—an issue in the science of ecology, which studies energy and resources. Ecologists are those who know best how questions of energy can be answered by our scientific knowledge. Finally, of course, it is also an *Economic* issue. But economics is also a matter bearing on health, quality of life, morality, and science. Energy can only be made sense of in terms of systems thinking, a form of thought ecologists specialize in.

The real American Way of Life is not just about making money and consumerism. It is a profoundly *Moral* way of life. The "pursuit of happiness" is a pursuit of *Quality of Life,* which has to do in part with a well-functioning economy, but it has just as much to do with health, with the aesthetics of everyday life, and with our visceral connection to nature. In the deepest sense, energy is a *Way of Life* issue. It is up to progressives to articulate it as such.

When progressives look closely at their own moral system—at the source of all their political, social, interpersonal, and even religious values—then the right way to frame the issues of the day becomes clearer.

RAISING CHILDREN

As we saw in Chapter 21, conservatives understand the centrality of the Strict Father family to their politics and their overall belief system. As a result, they spend well over 200 million dollars per year directly on programs for raising children—on promoting the Strict Father model of the family. They know well that as the twig is bent, so grows the tree. Liberals pay virtually no attention to the issue of how children are to be raised. They do not put any significant time or effort into what is most central in progressive thought—the Nurturant Parent model of the family. The issue is not even on the liberal radar screen.

The issue has many dimensions which liberals must address:

- *The effect of a Strict Father upbringing on individual children is all too often disastrous,* as we saw in Chapter 21. Allowing conservatives virtual free reign to give advice to parents on the raising of children is a national disaster—and liberals should be deeply ashamed for allowing it to happen. Remember that James Dobson alone is broadcast on over 1,600 radio stations. The nation's fathering programs (like The Promise Keepers) primarily promote the return of strict fathering.
- *The more children brought up with Strict Father values, the more future conservatives we will have.* That's why William Bennett writes books of children's moral stories based on conservative morality. If progressives want young Americans to grow up to have progressive values, the first thing they should do is promote Nurturant Parenting.
- *It is strict fathers (or mothers with those values) who are mostly responsible for spousal abuse and child abuse.* The most effective long-term way to lessen their abuses is to reduce the number of Strict Fathers, which means promoting Nurturant Parenting.
- *Right now decisions are being made about the content of moral education and standards for centers for early childhood learning.* Nurturant Parenting and its values should be explicitly put on the agenda. Conservatives have mostly had free reign to put forth their values as general American values, which they are not.

Unfortunately most progressives don't even want to talk about raising children. We can see why when we look at the five major groupings of progressives:

1. Socio-economic Progressives: Everything important in society and politics is a matter of economics and class—which is taken to include race, gender, ethnicity, and sexual preference.
2. Identity Politics Progressives: What is most important is redressing the real grievances of oppressed groups.
3. Greens: The central issues to be addressed are the ecology and the rights of indigenous peoples.
4. Civil Liberties Progressives: Individual rights and civil liberties are most important.
5. Anti-authoritarian Progressives: Fighting authoritarian practices and institutions—say, oppressive aspects of the state and of large corporations—is the uppermost concern.

Each major group focuses on real issues, but the most important issue of all—raising children—is left out. Nurturant Parenting is neither a socioeconomic question, nor an issue of identity politics, nor an ecological concern, nor a question of civil liberties, nor of authoritarian political institutions. It is not even on their radar screens. There are progressives primarily concerned with children and family issues, but such issues are child care, children's health, prenatal care for mothers, child poverty, education, the problems of minority children, child abuse, and so on. These very real concerns reflect the interests of the major groupings, but, to my knowledge, such groups are not principally concerned with the promotion of Nurturant Parenting itself.

All these forms of progressive politics place greatest interest in one dimension or another of progressive values. Each individual group focuses on its own individual issues, which results in a fragmentation of progressive politics. Conservatives have, to a significant extent, come together on the question of their central values. Progressives must also unite, if they are to have a chance of holding their own against the conservative juggernaut. I think the issue to bring progressives together should be this most central of all issues—raising children to become responsible, empathetic adults.

---

1. For more information on the financial support and dissemination of conservative research, see:

Covington, S. 1998. Moving a public policy agenda: The strategic philanthropy of conservative foundations. Report from the National Committee for Responsive Philanthropy.

Paget, K. 1998. Lessons of right-wing philanthropy. In *The American Prospect* 9: 89–95.

People for the American Way. 1996. *Buying a movement: Right-wing foundations and American politics.* Available at http://www.pfaw.org/issues/right/rw/reprwfound.html.

Stefancic, J., and R. Delgado. 1996. *No mercy: How conservative think tanks and foundations changed America's social agenda.* Philadelphia: Temple University Press.

Afterword, 2016

The Highest Conservative Principle

If there is one thing that must be maintained in a strict father family, it is the authority of the father and his dominant role in family life. When applied to politics, this Maintenance of Authority principle has a crucial correlate that often goes unnoticed by progressives and the news media. The highest conservative political principle is the Maintenance of Conservative Authority—the preservation, support, and extension of conservatism itself.

This principle explains something that progressives don't understand and consequently don't complain about. For example, President Obama often proposed policies that were originally conservative policies. A notable example is the Affordable Care Act, in which the taxpayers pay to get more customers for private insurance firms, customers who otherwise couldn't afford health care. This was originally proposed by a conservative think tank and put into practice in Massachusetts by Mitt Romney, a conservative Republican governor. But instead of thanking the president for accepting a conservative health care plan, conservatives have endlessly attacked "Obamacare" as a form of socialism, government overreach, government intrusion into private life where it does not belong, dictatorship, and even Nazism. Republican presidential candidates say it is the first thing they would get rid of. The conservative-led House of Representatives has voted over forty times to eliminate it. Why? Because, if it is successful, as it appears to be, it would give President Obama a victory and thus hurt the overall conservative cause. It would violate the principle of maintaining conservative authority above all else (see Chapter 9, p. 166).

The Private Depends on the Public

We have just seen in detail that conservative and progressive politics are based on opposed moral worldviews. In Chapter 20, I asked whether my own progressive political views were no more than an inherited ideology. I discussed there the important *non-ideological* reasons for my moral worldview. More recently, I have noticed another crucial factual reason for maintaining a progressive worldview.

From the beginning, the United States has been based on a nurturant principle:

- Citizens care about other citizens and work, through their government, to provide public resources for all.

Those public resources make both private enterprise and private life possible. From the beginning of the republic, transportation on public roads and bridges made business possible, as did the regulation of interstate commerce, the National Bank and the Patent Office, a judicial system that resolves business disputes, and the Post Office, which carried business mail. Private life was enriched by public education and public hospitals, as well as the roads and bridges and law enforcement. The Founding Fathers understood the importance of public resources for private life.

These days the private depends on the public in thousands of ways. Computer science was developed via grants from the National Science Foundation. The internet began as the Arpanet, developed by the Defense Department. Satellite technology was developed by NASA. Modern medicine and pharmaceuticals came about via NIMH. The GPS system was developed by the U.S. military, and hundreds of millions of Americans, including all cellphone users, now depend on it. The interstate highway system made trucking possible for business and expanded travel for private citizens. The Federal government helps support airports, and our Air Force trains most domestic pilots. The Federal Reserve makes money available for business. Not only do our state universities educate a great many of business's managerial and technological employees, but our private universities often receive more governmental money than do state universities. Other government agencies prevent disease and help provide the safe food, drinking water, and air required for private health and healthy employees in businesses. All of this is—in addition to huge business subsidies and tax breaks—paid for by the public.

Conservatives deny that the private depends on public resources. Citing Ronald Reagan's credo that "The Government is the Problem," conservatives are constantly trying to eliminate public resources and to privatize as much as possible. For example, they have proposed privatizing Social Security, replacing it with private investment in markets. Conservatives want to privatize public schools via vouchers and charter schools. Most charter schools are owned by private corporations, and they are no better than public schools. They use public funds (allotted per pupil) for private profit, generally paying teachers less and pocketing the difference. Conservatives have even proposed eliminating the Department of Education. The same is true for health care. Overturning "Obamacare" would increase the privatization of health care and eliminate government subsidies for those who cannot afford health care. Locally, conservatives have proposed, and sometimes succeeded in, privatizing public water supplies. And in the area of regulation, conservatives tend to favor eliminating government regulations in favor of private companies regulating themselves. For example, they have proposed eliminating the Environmental Protection Agency and the Food and Drug Administration, leaving environmental, food, and pharmaceutical regulation to private companies. They want to eliminate the Post Office in favor of private mailing services, refusing to let the Post Office extend services to seriously compete with private mail services. And they want to allow private corporations to frack on public lands, extracting oil and gas for their profit not the public's benefit, and to graze cattle cheaply or for free on public lands.

Private enterprise and private life depend on nurturant morality, but so does freedom in American life. Freedom is what public resources provide—freedom in a way that we take for granted but that needs to be brought out in the open. The freedom to start a business and to work in one—that is, the freedom to earn a living. The freedom of the body—to breathe air, drink water, and eat food that won't kill you, freedoms provided by the Food and Drug Administration, the Environmental Protection Agency, and the Centers for Disease Control. The Affordable Care Act has brought the freedom offered by health to millions of Americans. If you get cancer, or even just break your leg, and you have no health care, you will not be free. If you cannot earn enough to make a reasonable living from the best job you can get, you are not free. This is why increases in the minimum wage and in Social Security are freedom issues.

Although the words "freedom" and "liberty" are used regularly by conservative groups, conservatives mean something very different by these words. In Strict Father Morality applied to politics, "freedom" defined only by personal and not social responsibility means that one should be "free" to seek one's own profit and advantage without regard to how others—and the environment—might be harmed.

Freedom does not, or should not, mean that you are free to harm others. You may be free to walk down the street, but in doing so, you are not free to knock down others, push someone aside into an oncoming bus, or kick a passerby in the shins. You may be free to start a food business, but you are not free to put poisonous or cancer-causing ingredients in the food you sell.

How the Moral Order Limits Freedom

The Moral Order metaphor (Chapter 5, p. 104–5) links morality to authority: the most moral should rule. The father-knows-best assumption—that the strict father knows right from wrong—gives him authority. His wife is there to support his authority. And his children, who are born to do whatever feels good, need to be made into moral beings through punishment when they do wrong. It is through physical discipline that they learn moral discipline: to avoid punishment, they do what the father says. In this family model, the father's authority and power appropriately reflects his moral superiority, which comes from supporting and protecting the family while knowing right from wrong.

The Moral Order metaphor maps the authority–morality link onto other domains. The result is a hierarchy that is understood as just; power *should* rest with the most moral. Extending the hierarchy introduced in the first edition, we have the following:

> God over Human Beings; Human Beings over Nature; Adults over Children; The Rich over The Poor (and hence, Employers over Employees); Western Culture over Non-Western Culture; Our Country over Other Countries.

Most conservatives accept this much; others extend the hierarchy further:

> Men over Women; Whites over Nonwhites; Straights over Gays; Christians over Non-Christians.

In each case, the hierarchy limits the freedom of those lower on the hierarchy by legitimating the power of those higher on the hierarchy.

As we will see, a large proportion of conservative political policies follow from the conservative Moral Order. Remember, as we proceed, that the conservative Moral Order is a natural facet of Strict Father Morality applied to the full range of human experience. Let's look at these cases one at a time.

God over Human Beings

In conservative forms of religion, God is a Strict Father who, in effect, says: "Do as I say and you will get to Heaven; otherwise, you will burn in hell—punished painfully forever!" Who decides what God meant? A conservative Christian pastor is given that authority by his church. In progressive forms of religion, God is a Nurturant Parent who offers Grace (metaphorical nurturance: pp. 255–58), love, forgiveness, and fulfillment.

Conservatives have long made "freedom of religion" into a political issue, on the idea that the laws of God trump the laws of man. Traditional issues include school prayer (whether Bible verses should be read, or prayer required, in public schools), whether crosses can be erected, or Christmas scenes displayed using public money in public places, and whether Christian judges can post the Ten Commandments in public courtrooms. Kim Davis became a conservative hero when as a public clerk she cited her reading of the Bible in refusing to issue marriage licenses to gay couples in defiance of state law. Conservatives have proposed laws allowing owners of publicly licensed pharmacies to refuse to sell contraceptives to customers if their religion disapproves of contraception (e.g., if they are conservative Catholics). Catholic universities have cited freedom of religion in refusing to authorize birth control to women otherwise covered for it on their university health care plans. President George W. Bush's first executive order was that no foreign aid funds could go to providing birth control in countries where it was not available or affordable. And conservatives have made Planned Parenthood an issue, closing sites down in Texas and other states, based on the right-wing religious view that persons are created upon conception, that fertilized egg cells are "babies," and that therefore abortion is murder. The result has been that tens of thousands of women are being denied the freedom to control their own bodies and their own lives, even if they want to plan their parenthood, get contraception, and have children later.

Progressives have argued that the Constitution forbids the establishment of a state religion, citing Thomas Jefferson's invocation of a "solid wall" between religion and the state. But this formulation misses even deeper denials of religious freedom by conservatives—both the freedom to live by one's own religion, not someone else's, and the freedom to control your own body and your own life.

Human Beings over Nature

Conservatives constantly speak of nature as a gift from God to human beings, something for individual people (and corporations) to exploit and use for their own profit. Conservative commitment to individual (not social) responsibility and laissez-faire free markets where profit is maximized with little or no regulation clashes with most attempts to halt and reverse global warming, minimize pollution, protect endangered species, and preserve as much as possible of the natural world. Fish in rivers are being wiped out by the creation of dams for human recreation, the depletion of water for local farming, and the dumping of pollutants into streams by corporations who profit by not having to clean up their own mess. Coal, oil, and gas companies contribute to politicians, especially conservative politicians, to keep reaping profits at the cost of increasing global warming and threatening the future of the planet. Conservatives accordingly vote against subsidies or other legislation that would promote solar and wind energy. For forty years, conservatives have been primarily responsible for the

U.S. doing so little to preserve our planet and to protect against ongoing and future weather disasters caused by global warming.

Systemic Causation

The words "caused by global warming" have become fighting words. The reason is that there are two different notions of causation, and most people are working with only one of them.

Direct causation occurs when you use a force that results in an immediate local change, right then and there—like picking up a glass of water and taking a drink. In the strict father family the use of the father's authority and power is typically direct, say in immediate punishment of children for disobedience.

The world ecology is a system, and so has systemic causes, where it is the structure of the system that causes disasters to occur. There are four types of systemic causation: (1) chains of causation, (2) interacting causes, (3) feedback loops, and (4) probabilistic causes. For example, global warming can cause extreme, *cold* weather. Warming, for instance, can cause increased evaporation over the Pacific Ocean, which means more high-energy water molecules going into the air. This moisture is then blown north and east, over the North Pole, where the air is locally cold most of the year. There the moisture turns to snow, which is blown down over the East Coast, creating a great snowstorm in Washington, D.C. Conservatives say, "What global warming? This is more snow than we've ever seen!" They don't understand systemic causation. Nor does the average person.

Every language in the world has a way to express direct causation in its grammar. No language in the world has a way to express systemic causation in its grammar. Understanding direct causation comes naturally. Systemic causation must be taught if the effects of global warming are to be seriously understood.

The Rich over The Poor (Special Case: Employers over Employees)

In 1992, I sat in my living room watching Dan Quayle's acceptance speech for the vice-presidency. In it he gave a one-sentence argument against the progressive income tax—against taxing the rich at a higher rate than the poor. The sentence: "Why should the best people be punished?" There was cheering and a waving of signs!

It is now widely known that most of the wealth in this country has gone and is going to the top 1 percent, if not the top one-tenth of 1 percent. Conservatives have consistently voted to cut taxes on the rich. Current proposals for a "flat tax" would mean cutting taxes on the rich. The arguments are often versions of the Moral Hierarchy principle, that the rich are just better people than the poor: they have earned their money through hard work; the poor just haven't worked hard enough and so deserve to be poor. Hmmm . . . Corporate CEOs tend to make about two hundred times as much as their average employees—they can't be working two hundred times as hard. And many of the rich have inherited their money not earned it through work.

Another argument: The rich are job creators. They are helping the poor. If their taxes are cut, they can create more jobs. Ah, facts again. The rich actually tend to keep their money by buying prime real estate, yachts, art, and the like—things that will appreciate in value without giving back much to the economy as a whole. Moreover, the rich don't just give jobs out of the goodness of their hearts. Look at economics from the perspective of work: Working people are *profit-creators*. The rich only create jobs when they can get employees who can create profit for them. The poor create profit for the rich.

Conservatives have proposed cutting pension plans and benefits like contributions to health care. Think for a moment what pensions are: Pensions are delayed payments for work already done. If employees' pensions are cut, the company is stealing their money—money they have already earned. If the corporation says it can no longer pay "generous" benefits, then the company is cutting employees' salaries. "Benefits" are not gifts; "generosity" is not at issue. Benefits are part of pay for work.

Corporations now tend to think of their employees as being either "assets" or "resources." The "assets" are seen as special people: top managers or important "creative" people or technical specialists who are hard to find and hence valuable. The "resources" are interchangeable, fitting a job description, and hired through the "human resources" department. They are treated precisely as resources. One makes the most profit by minimizing the use and cost of resources: paying as little as possible to resource-workers, letting as many as possible go, replacing them with technology whenever feasible, and outsourcing whenever it is profitable. Many people are poor because they have lost jobs or have jobs that don't pay a living wage. Conservatives see them as not trying hard enough to find work or pull themselves up by their bootstraps. Many people are poor because they lack education, or because they have accumulated too much debt getting an education. Conservatives thus want to cut unemployment insurance, not allow an increase in the minimum wage, and cut funds for education. These are all examples of conservatives seeing the rich as more moral than the poor—the rich deserve to be rich; the poor deserve to be poor. It follows from Strict Father Morality.

Conservatives are also against unions and want to legislate them out of existence via what are called "right to work" laws. Such laws see employment through a Strict Father lens: as simply a matter of individual responsibility by the employee. Conservative enmity against unions follows from the moral hierarchy: Rich Over Poor; Employer Over Employee. Unions are actually agents of freedom—freedom from corporate servitude and wage slavery. Without unions, employees have to individually take what is offered, usually far less than they would get with a union: not just pay but worker safety, health care benefits, pensions, reasonable working conditions and hours, reasonable vacation time. What is "reasonable"? What the union members can negotiate. Unions create freedom.

Austerity

The idea that the rich are more moral than the poor extends to nations. The rich European nations of the north are seen as more moral than the poorer European nations to the south, like Greece, Portugal, and Spain. The northern Europeans assume that southern European workers are just lazy, deserve their suffering, and should "tighten their belts." What is needed instead is wise investment in the south by northern banks and corporations—investments in infrastructure building, education, and business that can gainfully employ unemployed workers who are already educated. In Spain, for example, educated Spaniards are leaving in droves for jobs elsewhere in Europe that fit their educational level.

Adults Over Children

There is a basic fact about children's brain development that needs to be widely known. By the time a child is about five years old, half of her neural connections have died off—the half that is least used. That has a consequence. If children are to find fulfillment in life, they need to start earlier than five. They need early childhood education—not high-powered stuff, but a wide basic introduction to many areas—some basic math, reading and writing, art, music, a foreign language, movement or sports. Five-year-olds can do more than many people realize.

Conservatives tend not to want to fund early childhood education. First, they see it as a government program to be destroyed. Second, they are concerned about what would be taught—perhaps not the subject matter of what would be taught in a strict father household. They want to be sure that children are taught a conservative way of looking at the world.

Children learn best in nurturant situations, when they are cared for and their needs as individuals are taken seriously. This requires small classroom sizes and teachers' aides. The best teachers are nurturers. Conservatives believe that strictness is the only way to teach. Nurturant teachers do not provide a version of the strict parent environment. The debate about teaching to the test and punishing "failing schools" leaves out the nurturant dimension of education.

In a strict father family, it not merely allowed for parents to beat children when they don't do as told or show disrespect; it is *required* that the child be painfully punished; it is a parental duty. Since the father knows what is moral and the mother must be supportive of the father, a disobedient child is both violating the authority of the father and will not grow up to be a moral being. Corporal punishment for disobedience used to be normal, both at home and in school. In schools, it has now been banned in thirty-one states in the U.S. But corporal punishment—beating, often with sticks or "paddles"—is still allowed in nineteen states: Alabama, Arizona, Arkansas, Colorado, Florida, Georgia, Idaho, Indiana, Kansas, Kentucky, Louisiana, Mississippi, Missouri, North Carolina, Oklahoma, South Carolina, Tennessee, Texas, and Wyoming. These are all states where conservatism is dominant.

Western Culture over Non-Western Culture

This assumption in the Moral Hierarchy has been realized in foreign policy assumptions about the superiority of capitalism and democratic elections. It has been widely assumed that the problems in non-Western countries could be solved by overthrowing foreign dictatorships in poor countries, often by force, and introducing elections and global capitalism dominated by Western corporations. This was the idea behind the wars in Iraq and Afghanistan, as well as Western support for the Arab Spring revolts in Egypt, Libya, and Syria. The Western ideals of democracy and global capitalism were supposed to quickly show their advantages over non-Western traditions: rule by religion, tribalism (with tribal hatreds and warlords), jihadism, and institutionalized corruption. The fall of communism and the breakup of the Soviet Union was supposed to lead to democracy and Western-style capitalism, even when there were no prior democratic or capitalist institutions in place. In China, the introduction of capitalism was supposed to lead to democracy. It didn't.

Though the moral hierarchy is a natural product of Strict Father Morality and a conservative worldview, this view about the superiority of Western culture has been held by conservative and liberal administrations alike. However, American foreign policy institutions like the CIA, the NSA, and the military share a strong Strict Father culture seen in "realist" foreign policy as opposed to "soft power" foreign policy. Foreign policy has not been looked at from the perspective of the Strict Father moral hierarchy. It is an open question as to whether such a perspective would be helpful. My job here is to raise the issue.

Our Nation over Other Nations

There is a big difference between patriotism and jingoism, which is often mistaken for patriotism. Patriotism is about improving the lives of one's fellow citizens and improving one's country's contribution to the world. In the conservative moral hierarchy, our country is taken as simply better than other countries. This is jingoism, not true patriotism, which rests on progressive values. Democratic candidates in the 2016 election campaign have chosen patriotism over jingoism, focusing on conservative failures for the American people, while bringing up successes for the citizens of other wealthy countries where America has failed: universal healthcare as a right, free college education as a national investment, family leave, higher wage scales, much greater equality of wealth, safer gun laws, and better environmental protections. Each of these is a freedom issue: you are just not free if you lack health care and get seriously injured or ill; if you lack an education; if you have to choose between keeping your job and caring for your child; if you work but don't make a wage you can live on; if you are threatened by gun violence; if the planet you live on is being destroyed and you are threatened by climate disasters.

Looking at successes that have occurred in other countries and asking why we cannot achieve them here in the U.S. is a patriotic response to the jingoism of assuming that our nation is simply more moral than other nations.

At this point we turn to some of the nastier parts of the moral hierarchy—forms of discrimination: sexism, racism, homophobia, and religious discrimination. It is by no means the case that all conservatives share these forms of discrimination and that they are absent in all liberals. But these parts of the moral hierarchy do show up in conservative politics.

Men over Women

The "war on women" framing by liberal feminists reflects their experience that women are being harmed via purposeful acts by conservatives. But conservatives don't see themselves as conducting such a war. Instead they see themselves as simply acting morally according to the conservative moral worldview. In the strict father family, it is the father who determines what the women in the family should and should not do, sexually, reproductively, and with their bodies. Accordingly, men are above women in the conservative moral hierarchy. And a great many conservative women share the conservative moral worldview with conservative men and see no "war on women." Male dominance in the strict father family is real, and it translates into male superiority in conservative politics.

As we have seen, the conservative moral hierarchy contradicts progressive views of freedom. The freedom at issue here is the freedom to control your own body. The decision to engage in sex is one such decision. The decision not to have a child now—or at all, or not with a given sex partner—is also such a decision. To exercise such decisions, a woman needs sex education and the availability of contraception or morning-after pills. These are methods to avoid an unwanted pregnancy by exerting control over your own body.

In a strict father family, it is typically the father who makes decisions as to child-bearing—either by his wife or by a daughter. Hence, conservatives have introduced legislation in a number of states requiring parental notification or spousal notification in case a woman wants to end a pregnancy. Such laws put control in the hands of the father.

An unwanted pregnancy can occur in various stages. A fertilized egg first develops into a group of cells called a blastocyst, which only later can be embedded into a uterus and become an embryo, which does not yet have human form. The embryo begins to develop into a fetus between eight and twelve weeks. These are all parts of a woman's body and can affect a woman's health. A pregnancy can be ended by an operation at any of these stages. Such an operation, at any stage from blastocyst on, is called an "abortion."

The Planned Parenthood organization was founded to allow women to make and follow through on informed decisions on such matters. It provides diagnosis and care for sexually-transmitted diseases, sex education materials and contraceptives, as well as morning-after pills. It also provides abortions, mostly for women who cannot otherwise afford them.

Throughout history women have attempted to control their own bodies by avoiding or ending pregnancies. For many years, abortions were legally banned in many states but were nonetheless performed. Wealthy women could travel to states or other countries to have safe abortions. But women who could not afford such trips had to rely on "back-alley abortions"— unsafe illegal abortions that could leave a woman without a later ability to have a child, or with terrible internal injuries, or even death. The Supreme Court's 1973 decision in *Roe vs. Wade* made abortions legal under very general conditions, but this freedom is being slowly eroded by subsequent decisions by Supreme Courts with more conservative majorities.

Over several years, the Texas legislature gradually placed more and more restrictions on women's health centers performing safe abortions for impoverished women until hardly any were left. In October 2015, Texas barred Planned Parenthood from receiving state Medicaid money, making it impossible for an impoverished woman to obtain a safe legal operation to end a pregnancy in the entire state of Texas.

Another imposition on the freedom of a woman to control her own body is the forced ultrasound, imposed by conservative legislatures. Thirteen states require that women seeking an abortion must have a (medically unnecessary) ultrasound, a thirty-minute procedure that varies from uncomfortable to painful, especially if you are pregnant. The woman is then forced to view the ultrasound picture before the operation—clearly an attempt to have her change her mind. Similarly when an ultrasound is performed as preparation for an abortion procedure, fifteen other states require that the woman must be offered the opportunity to view the ultrasound picture.

There are of course other issues surrounding this aspect of the conservative moral hierarchy. Conservatives have consistently kept the Equal Rights Amendment from being passed, as well as other laws guaranteeing equal pay for equal work and family leave.

Whites over Nonwhites

The Black Lives Matter movement has arisen for a reason. Innocent black men were being killed, as well as arrested and assaulted, because whites, especially police, assumed that most blacks, especially young black men, are up to no good. Throughout America, police are performing stop-and-frisk operations on young black men and pulling over drivers for the offense of driving while black. Moreover, America is filling its prisons with young black men convicted only of minor drug offenses and not dangerous to others. Conservatives have passed "Stand Your Ground" laws that permit shooting people deemed threatening, and sadly there are whites who seem to find the mere presence of blacks threatening.

This part of the conservative moral hierarchy also arises in the debate over immigration. It boiled over in 2015 when Republican presidential candidate Donald Trump called Mexicans immigrants "murderers and rapists." Conservatives often call those entering the U.S. without papers but otherwise living normal law-abiding lives "illegals", as if they were career criminals, or "aliens," as if they are from another planet. Mostly nonwhite, immigrants are considered by many conservatives to be immoral, and hence as people who should be deported. Conservatives have also been against providing health care and education to children of undocumented immigrants born and raised in the U.S. who have full American citizenship.

Straights over Gays

Previous bans on same-sex marriages were pushed through by conservatives. Conservatives in many states are still trying to find ways to allow country clerks to refuse to grant marriage licenses to same-sex couples. Conservative Christians still consider homosexuality an abomination and want marriage to be legally defined as between a man and a woman.

Christian over Non-Christians

Conservative Christians have provided the principal support in the polls for Ben Carson as President. Carson, in 2015, remarked that he considered it unconstitutional for a Muslim to become President. He sees Christians as morally superior, as do those who continued to support him after that remark. It is commonplace for conservatives to claim that the U.S. is a

Christian country and to proclaim their Christianity as a qualification for public office.

* * *

By now you can see that the positions on issues in political campaigns are not a random list. Conservative principles fit together because they follow from Strict Father morality. Conservative positions are logical from this perspective; they make sense, just as progressive positions make sense from the perspective of Nurturant Parent morality.

As we pointed out, the common principles of conservative thought make up a list crying out for an explanation of how they fit together: smaller government; strong defense; lower taxes; traditional family values; personal responsibility; and free markets. As we have seen, these all flow from a strict father morality.

Finally, let me close this edition with responses to two basic questions that perplex liberals.

1. Why should poor conservatives vote against their financial interests?

    Because they are voting their moral identities, not their pocketbooks. They are voting for people who believe in what they believe, and they want to see a world in which their moral principles are upheld

2. Why do Tea Party members of Congress obstruct even the workings of an overwhelmingly conservative Congress?

    Because they believe that compromising with progressive positions is immoral on the grounds that it weakens and undermines the authority of conservatism. It would be like a strict father giving in and compromising his authority in the family.

* * *

Our deepest moral worldviews are unconscious. That should not be surprising. After all, all thought is physical and works via neural circuitry. We do not, and cannot, have immediate direct access to our deepest modes of thought. It takes serious study in cognitive science to figure out the details of our moral worldviews. But once you see them, you see that they are real and that they explain a lot.

They are also remarkably stable. Despite all that has changed since 1996, the basic worldviews and their logics are still very much present. If you follow politics, you encounter them every day—and often on totally different issues.

References

This is a topic-oriented list of references. It includes both works cited and other works that are either of an introductory or supplementary nature. It is intended to allow the reader entry to the literature, rather than to be exhaustive.

References in the text refer to the letters and numbers that structure this list. The list of categories appears first, then the category-by-category references.

Organization

A. Cognitive Science and Cognitive Linguistics
   1. Metaphor Theory
   2. Categorization
   3. Framing
   4. Discourse and Pragmatics
   5. Decision Theory: The Heuristics and Biases Approach
   6. Cognitive Science and Moral Theory

B. Child Development and Childrearing
   1. Attachment Research
   2. Socialization Research
   3. Fundamentalist Christian Childrearing Manuals

4. Mainstream Childrearing Manuals
5. Childrearing and Violence
6. Critiques of Fundamentalist Childrearing
7. Background: Childrearing and National Character

C. Politics
1. Conservative Political Writings
2. Neoconservatism
3. Modern Theoretical Liberalism
4. Communitarian Critiques
5. Modern Theoretical Libertarianism

D. Public Administration
1. Bureaucratic Reform
2. Stars Wars Policy

E. Miscellaneous

A. Cognitive Science and Cognitive Linguistics

The Baumgartner-Payr and Solso-Massaro books provide some sense of the range of questions that cognitive scientists consider. Edelman's book contains not only an overview of his own work but an account of how cognitive linguistics meshes with research in neuroscience.

Baumgartner, P., and S. Payr. 1995. *Speaking minds: Interviews with twenty eminent cognitive scientists*. Princeton: Princeton University Press.

Edelman, G. M. 1992. *Bright air, brilliant fire: On the matter of the mind*. New York: Basic Books.

Solso, R. L., and D. W. Massaro. 1995. *The science of the mind: 2001 and beyond*. New York: Oxford University Press.

A1. Metaphor Theory

The most popular introduction to the field is the Lakoff-Johnson book. Lakoff 1993 is the most recent general survey of the field. The journal *Metaphor and Symbolic Activity* is devoted primarily to empirical psychological research on metaphor. Gibbs 1994 is an excellent overview of that research. Johnson 1981 is a survey of previous approaches to the study of metaphor. *Cognitive Linguistics* is a more general journal devoted not only to metaphor research but to the whole gamut of cognitive approaches to linguistics.

Gentner, D., and D. R. Gentner. 1982. Flowing waters or teeming crowds: Mental models of electricity. In *Mental models,* edited by D. Gentner and A. L. Stevens. Hillsdale, N.J.: Erlbaum.

Gibbs, R. 1994. *The poetics of mind: Figurative thought, language, and understanding.* Cambridge: Cambridge University Press.

Johnson, M. 1987. *The body in the mind: The bodily basis of meaning, imagination, and reason.* Chicago: University of Chicago Press.

Johnson, M., ed. 1981. *Philosophical perspectives on metaphor.* Minneapolis: University of Minnesota Press.

Klingebiel, C. 1990. The bottom line in moral accounting. Manuscript, University of California, Berkeley.

Lakoff, G. 1993. The contemporary theory of metaphor. In *Metaphor and thought,* 2d ed., edited by A. Ortony, 202–51. Cambridge: Cambridge University Press.

Lakoff, G., and M. Johnson. 1980. *Metaphors we live by.* Chicago: University of Chicago Press.

Lakoff, G., and M. Turner. 1989. *More than cool reason: A field guide to poetic metaphor.* Chicago: University of Chicago Press.

Reddy, M. 1979. The conduit metaphor. In *Metaphor and thought,* edited by A. Ortony, 284–324. Cambridge: Cambridge University Press.

Sweetser, E. (In preparation). Our Father, our king: What makes a good metaphor for God.

Sweetser, E. 1990. *From etymology to pragmatics: Metaphorical and cultural aspects of semantic structure.* Cambridge: Cambridge University Press.

Taub, S. 1990. Moral accounting. Manuscript, University of California, Berkeley.

Turner, M. 1991. *Reading minds: The study of English in the age of cognitive science.* Princeton: Princeton University Press.

Winter, S. 1989. Transcendental nonsense, metaphoric reasoning and the cognitive stakes for law. *University of Pennsylvania Law Review* 137, 1105–1237.

A2. Categorization

Lakoff 1987 is a survey of categorization research up to the mid-1980s. The papers by Rosch are the classics in prototype theory.

Barsalou, L. W. 1983. Ad-hoc categories. *Memory and Cognition* 11:211–27.

Barsalou, L. W. 1984. Determination of graded structures in categories. Psychology Department, Emory University, Atlanta.

Brugman, C. 1988. *Story of "over."* New York: Garland.

Kay, P. 1983. Linguistic competence and folk theories of language: Two English hedges. In *Proceedings of the Ninth Annual Meeting of the Berkeley Linguistics Society*, 128–37. Berkeley: Berkeley Linguistics Society.

Kay, P., and C. McDaniel. 1978. The linguistic significance of the meanings of basic color terms. *Language* 54:610–46.

Lakoff, G. 1972. Hedges: A study in meaning criteria and the logic of fuzzy concepts. In *Papers from the Eighth Regional Meeting, Chicago Linguistic Society*, 183–228. Chicago: Chicago Linguistic Society. Reprinted in *Journal of Philosophical Logic* 2 (1973):458–508.

Lakoff, G. 1987. *Women, fire, and dangerous things: What categories reveal about the mind.* Chicago: University of Chicago Press.

McNeill, D., and P. Freiberger. 1993. *Fuzzy logic.* New York: Simon and Schuster.

Rosch, E. (E. Heider). 1973. Natural categories. *Cognitive Psychology* 4:328–50.

Rosch, E. 1975a. Cognitive reference points. *Cognitive Psychology* 7:532–47.

Rosch, E. 1975b. Cognitive representations of semantic categories. *Journal of Experimental Psychology: General* 104:192–233.

Rosch, E. 1977. Human categorization. In *Studies in cross-cultural psychology*, edited by N. Warren. London: Academic Press.

Rosch, E. 1978. Principles of categorization. In *Cognition and categorization*, edited by E. Rosch and B. B. Lloyd, 27–48. Hillsdale, N.J.: Erlbaum.

Rosch, E. 1981. Prototype classification and logical classification: The two systems. In *New trends in cognitive representation: Challenges to Piaget's theory*, edited by E. Scholnick, 73–86. Hillsdale, N.J.: Erlbaum.

Rosch, E., and B. B. Lloyd. 1978. *Cognition and categorization.* Hillsdale, N.J.: Erlbaum.

Schwartz, A. 1992. Contested concepts in cognitive social science. Honors thesis, University of California, Berkeley.

Smith, E. E., and D. L. Medin. 1981. *Categories and concepts.* Cambridge: Harvard University Press.

Taylor, J. 1989. *Linguistic categorization: Prototypes in linguistic theory.* Oxford: Clarendon Press.

Wittgenstein, L. 1953. *Philosophical investigations.* New York: Macmillan.

Zadeh, L. 1965. Fuzzy sets. *Information and Control* 8:338–53.

A3. Framing

Fillmore is the major source for empirical linguistic research. Schank and Abelson started the major artificial intelligence approach. Holland and Quinn introduced the techniques to anthropology.

Fillmore, C. 1975. An alternative to checklist theories of meaning. In *Proceedings of the First Annual Meeting of the Berkeley Linguistics Society,* 123–31. Berkeley: Berkeley Linguistics Society.

Fillmore, C. 1976. Topics in lexical semantics. In *Current issues in linguistic theory,* edited by P. Cole, 76–138. Bloomington: Indiana University Press.

Fillmore, C. 1978. The organization of semantic information in the lexicon. In *Papers from the Parasession on the Lexicon,* 1–11. Chicago: Chicago Linguistic Society.

Fillmore, C. 1982a. Towards a descriptive framework for spatial deixis. In *Speech, place, and action,* edited by R. J. Jarvella and W. Klein, 31–59. London: Wiley.

Fillmore, C. 1982b. Frame semantics. In *Linguistics in the morning calm,* edited by the Linguistic Society of Korea, 111–38. Seoul: Hanshin.

Fillmore, C. 1985. Frames and the semantics of understanding. *Quaderni di Semantica* 6:222–53.

Holland, D. C., and N. Quinn, eds. 1987. *Cultural models in language and thought.* Cambridge: Cambridge University Press.

Schank, R. C., and R. P. Abelson. 1977. *Scripts, Plans, Goals, and Understanding.* Hillsdale, N.J.: Erlbaum.

A4. Discourse and Pragmatics

Green and Levinson are excellent introductory pragmatics texts. Schiffrin and the Brown-Yule book provide excellent ways into the discourse literature.

Brown, G., and G. Yule. 1983. *Discourse analysis*. Cambridge: Cambridge University Press.

Brown, P., and S. C. Levinson. 1987. *Politeness: Some universals in language usage*. Cambridge: Cambridge University Press.

Goffman, E. 1981. *Forms of talk*. Oxford: Basil Blackwell.

Gordon, D., and G. Lakoff. 1975. Conversational postulates. In *Syntax and semantics 3: Speech acts,* edited by P. Cole and J. L. Morgan, 83–106. New York: Academic Press.

Green, G. 1989. *Pragmatics and natural language understanding*. Hillsdale, N.J.: Erlbaum.

Grice, P. 1989. *Studies in the way of words*. Cambridge: Harvard University Press.

Gumperz, J. J. 1982a. *Discourse strategies*. Cambridge: Cambridge University Press.

Gumperz, J. J. 1982b. *Language and social identity*. Cambridge: Cambridge University Press.

Hall, E. T. 1976/1981. *Beyond culture*. New York: Anchor/Doubleday.

Keenan, E. O. 1976. The universality of conversational implicature. *Language in Society* 5:67–80.

Lakoff, R. 1973. The logic of politeness; or, minding your P's and Q's. In *Papers from the Ninth Regional Meeting of the Chicago Linguistic Society,* 292–305. Chicago: Chicago Linguistic Society.

Levinson, S. C. 1983. *Pragmatics*. Cambridge: Cambridge University Press.

Saville-Troike, M. 1989. *The ethnography of communication: An introduction*. 2d ed. Oxford: Basil Blackwell.

Schiffrin, D. 1994. *Approaches to discourse analysis*. Oxford: Basil Blackwell.

Scollon, R., and S. W. Scollon. 1995. *Intercultural communication: A discourse approach*. Oxford: Basil Blackwell.

Stubbs, M. 1983. *Discourse analysis: The sociolinguistic analysis of natural language*. Chicago: University of Chicago Press.

Tannen, D. 1986. *That's not what I meant!: How conversational style makes or breaks your relations with others.* New York: Morrow.

Tannen, D. 1991. *You just don't understand: Women and men in conversation.* New York: Ballantine.

Tannen, D., ed. 1993. *Framing in discourse.* New York: Oxford University Press.

van Dijk, T. 1985. *Handbook of discourse analysis.* New York and London: Academic Press.

Weiser, A. 1974. Deliberate ambiguity. In *Papers from the Tenth Regional Meeting of the Chicago Linguistic Society,* 723–31. Chicago: Chicago Linguistic Society.

Weiser, A. 1975. How not to answer a question: Purposive devices in conversational strategy. In *Papers from the Eleventh Regional Meeting of the Chicago Linguistic Society,* 649–60. Chicago: Chicago Linguistic Society.

A5. Decision Theory: The Heuristics and Biases Approach

These are sample papers from a huge literature. They are chosen largely because they demonstrate framing effects.

Kahneman, D., and A. Tversky. 1983. Can irrationality be intelligently discussed? *Behavioral and Brain Sciences* 6:509–10.

Kahneman, D., and A. Tversky. 1984. Choices, values, and frames. *American Psychologist,* 39:341–50.

Tversky, A., and D. Kahneman. 1974. Judgment under uncertainty: Heuristics and biases. *Science* 185:1124–31.

Tversky, A., and D. Kahneman. 1981. The framing of decisions and the psychology of choice. *Science* 211:453–58.

Tversky, A., and D. Kahneman. 1988. Rational choice and the framing of decisions. In *Decision making: Descriptive, normative, and prescriptive interactions,* edited by D. E. Bell, H. Raiffa, and A. Tversky, 167–92. Cambridge: Cambridge University Press.

A6. Cognitive Science and Moral Theory

Classical moral theory assumed that the empirical study of the mind could not affect moral issues. Three books by distinguished philosophers challenge that assumption.

Churchland, P. M. 1995. *The engine of reason, the seat of the soul: A philosophical journey into the brain.* Cambridge, Mass.: MIT Press.

Flanagan, O. 1991. *Varieties of moral personality: Ethics and psychological realism.* Cambridge: Harvard University Press.

Johnson, M. 1993. *Moral imagination: Implications of cognitive science for ethics.* Chicago: University of Chicago Press.

B. Child Development and Childrearing

B1. Attachment Research

Karen's book is the best overall introduction. I suggest you start there.

Ainsworth, M. D. S. 1967. *Infancy in Uganda: Infant care and the growth of love.* Baltimore: The Johns Hopkins University Press.

Ainsworth, M. D. S. 1969. Object relations, dependency and attachment: A theoretical view of the infant-mother relationship. *Child Development* 40:969–1025.

Ainsworth, M. D. S. 1983. A sketch of a career. In *Models of achievement: Reflections of eminent women in psychology,* edited by A. N. O'Connell and N. F. Russo, 200–219. New York: Columbia University Press.

Ainsworth, M. D. S. 1984. Attachment. In *Personality and the behavioral disorders,* edited by N. S. Endler and J. McV. Hunt, 1:559–602. New York: Wiley.

Ainsworth, M. D. S. 1985. Attachments across the lifespan. *Bulletin of the New York Academy of Medicine* 61:792–812.

Ainsworth, M. D. S., M. D. Blehar, E. Waters, and S. Wall. 1978. *Patterns of attachment: A psychological study of the strange situation.* Hillsdale, N.J.: Erlbaum.

Belsky, J., and J. Cassidy. (In press). Attachment: Theory and evidence. In *Developmental principles and clinical issues in psychology and psychiatry,* edited by M. Rutter, D. Hay, and S. Baron-Cohen. Oxford: Basil Blackwell.

Belsky, J., and L. V. Steinberg. 1978. The effects of day care: A critical review. *Child Development* 49:929–49.

Belsky, J., L. Youngblood, and E. Pensky. 1990. Childrearing history, marital quality, and maternal affect: Intergenerational transmission in a low-risk sample. *Development and Psychopathology* 1:291–304.

Bowlby, J. 1944. Forty-four juvenile thieves: Their characters and home life. *International Journal of Psycho-Analysis* 25:19–52, 107–27. Reprinted (1946) as monograph. London: Bailiere, Tindall and Cox.

Bowlby, J. 1951. Maternal care and health care. Geneva: *World Health Organization Monograph Series* 2.

Bowlby, J. 1958. The nature of the child's tie to his mother. *International Journal of Psycho-Analysis* 39:350–73.

Bowlby, J. 1967. Foreword. In M. D. S. Ainsworth, *Infancy in Uganda*. Baltimore: The Johns Hopkins University Press.

Bowlby, J. 1970. *Child care and the growth of love*. 2d ed. Harmondsworth, Middlesex: Penguin.

Bowlby, J. 1973. *Attachment and loss*. Vol. 2: *Separation*. New York: Basic Books.

Bowlby, J. 1979. *The making and breaking of affectional bonds*. New York: Routledge.

Bowlby, J. 1980. *Attachment and loss*. Vol. 3: *Loss, sadness and depression*. New York: Basic Books.

Bowlby, J. 1982. *Attachment and loss*. Vol. 1: *Attachment*. Rev. ed. New York: Basic Books.

Bowlby, J. 1988. *A secure base: Clinical applications of attachment theory*. London: Routledge.

Bowlby, J., K. Figlio, and R. Young. 1990. An interview with John Bowlby on the origins and reception of his work. *Free Associations* 21:36–64.

Brazelton, T. B. 1983. *Infants and mothers: Differences in development*. Rev. ed. New York: Delta.

Brazelton, T. B., B. Koslowski, and M. Main. 1974. The origins of reciprocity: The early mother-input interaction. In *The effect of the infant on its caregiver*, edited by M. Lewis and L. Rosenblum. New York: Wiley.

Goldberg, S. 1991. Recent developments in attachment theory. *Canadian Journal of Psychiatry* 36:393–400.

Karen, R. 1994. *Becoming attached: Unfolding the mystery of the infant-mother bond and its impact on later life*. New York: Warner Books.

Lieberman, A. F. 1993. *The emotional life of the toddler*. New York: The Free Press.

Main, M. 1991. Metacognitive knowledge, metacognitive monitoring, and singular (coherent) versus multiple (incoherent) model of attachment: Findings and directions for future research. In *Attachment across the life cycle*, edited by C. M. Parkes, J. Stevenson-Hinde, and P. Marris. New York: Tavistock/Routledge.

Main, M., and D. Weston. 1981. The quality of the toddler's relationship to mother and to father as related to conflict behavior and readiness to establish new relationships. *Child Development* 52: 932–40.

Main, M., and D. Weston. 1982. Avoidance of the attachment figure in infancy: Descriptions and interpretations. In *The place of attachment in human behavior*, edited by C. M. Parkes and J. Stevenson-Hinde, 31–59. New York: Basic Books.

Main, M., and E. Hesse. 1990. Parents' unresolved traumatic experiences are related to infant disorganized attachment status: Is frightened and/or frightening parental behavior the linking mechanism? In *Attachment in the preschool years*, edited by M. Greenberg, D. Cicchetti, and E. M. Cummings. Chicago: University of Chicago Press.

Main, M., N. Kaplan, and J. Cassidy. 1985. Security in infancy, childhood, and adulthood: A move to the level of representation. In *Growing points in attachment theory and research. Monographs of the Society for Research in Child Development* 50 (Serial No. 209), edited by I. Bretherton and E. Waters, 66–104.

Sroufe, L. A. 1979. Socioemotional development: A developmental perspective on day care. In *Handbook of infant development*, edited by J. Osofsky. New York: Wiley.

Sroufe, L. A., and E. Waters. 1977. Attachment as an organizational construct. *Child Development* 48:1184–89.

Sroufe, L. A., B. Egeland, and T. Kreutzer. 1990. The fate of early experience following developmental change: longitudinal approaches to individual adaptation in childhood. *Child Development* 61:1363–73.

Sroufe, L. A., N. E. Fox, and V. R. Pancake. 1983. Attachment and dependency in developmental perspective. *Child Development* 54:1615–27.

Sroufe, L. A., R. G. Cooper, and G. B. DeHart. 1992. *Child development: Its nature and course.* 2d ed. New York: McGraw-Hill.

Stern, D. N. 1985. *The interpersonal world of the infant.* New York: Basic Books.

Stern, D. 1977. *The first relationship*. Cambridge: Harvard University Press.

B2. Socialization Research

Maccoby and Martin is the best overall survey, though it stops in 1983. Baumrind 1991 covers research on authoritarian vs. authoritative childrearing up to 1991.

Apolonio, F. J. 1975. Preadolescents' self-esteem, sharing behavior, and perceptions of parental behavior. *Dissertation Abstracts* 35: 3406B.

Baldwin, A. L. 1948. Socialization and the parent-child relationship. *Child Development* 19:127–136.

Baldwin, A. L. 1949. The effect of home environment on nursery school behavior. *Child Development* 20:49–62.

Baldwin, A. L., J. Kalhoun, and F. H. Breese. 1945. Patterns of parent behavior. *Psychological Monographs* 58.

Baumrind, D. 1967. Child care practices anteceding 3 patterns of preschool behavior. *Genetic Psychology Monographs* 75:43–88.

Baumrind, D. 1971. Current patterns of parental authority. *Developmental Psychology Monograph* 4.

Baumrind, D. 1972. An exploratory study of socialization effects on black children: Some black-white comparisons. *Child Development* 43:261–67.

Baumrind, D. 1977. Socialization determinants of personal agency. Paper presented at the meeting of the Society for Research in Child Development, New Orleans, March 27–30.

Baumrind, D. 1979. Sex-related socialization effects. Paper presented at the meeting of the Society for Research in Child Development, San Francisco.

Baumrind, D. 1987. A developmental perspective on adolescent risk-taking behavior in contemporary America. In *New directions for child development: Adolescent health and social behavior*, edited by W. Damon, vol. 37: 92–126. San Francisco: Jossey-Bass.

Baumrind, D. 1989. Rearing competent children. In *Child development today and tomorrow*, edited by W. Damon, 349–78. San Francisco: Jossey-Bass.

Baumrind, D. 1991. Parenting styles and adolescent development. In *The encyclopedia of adolescence*, edited by R. Lerner, A. C. Petersen, and J. Brooks-Gunn, 746–58. New York: Garland.

Baumrind, D., and A. E. Black. 1967. Socialization practices associated with dimensions of competence in preschool boys and girls. *Child Development* 38:291–327.

Becker, W. C., D. R. Peterson, Z. Luria, D. J. Shoemaker, and L. A. Hellmer. 1962. Relations of factors derived from parent-interview ratings to behavior problems of five-year-olds. *Child Development* 33:509–35.

Block, J. 1971. *Lives through time*. Berkeley: Bancroft Books.

Burton, R. V. 1976. Honesty and dishonesty. In *Moral development and behavior*, edited by T. Lickona. New York: Holt, Rinehart and Winston.

Comstock, M. L. 1973. Effects of perceived parental behavior on self-esteem and adjustment. *Dissertation Abstracts* 34:465B.

Coopersmith, S. 1967. *The antecedents of self-esteem*. San Francisco: W. H. Freeman.

Egeland, B., and L. A. Sroufe. 1981a. Attachment and early maltreatment. *Child Development* 52:44–52.

Egeland, B., and L. A. Sroufe. 1981b. Developmental sequelae of maltreatment in infancy. *New Directions for Child Development* 11:77–92.

Eisenberg, N., and P. H. Mussen. 1989. *The roots of prosocial behavior in children*. Cambridge: Cambridge University Press.

Eron, L. D., L. O. Walder, and M. M. Lefkowitz. 1971. *Learning of aggression in children*. Boston: Little, Brown.

Feshbach, N. D. 1974. The relationship of child-rearing factors to children's aggression, empathy and related positive and negative social behaviors. In *Determinants and origins of aggressive behavior*, edited by J. deWitt and W. W. Hartup. The Hague: Mouton.

Gilligan, C. 1982. *In a different voice: Psychological theory and women's development*. Cambridge: Harvard University Press.

Gilligan, C. 1987. Adolescent development reconsidered. In *New directions for child development*, edited by C. E. Irwin, Jr., vol. 37:63–92. San Francisco: Jossey-Bass.

Goldstein, A. P., and G. Y. Michaels. 1985. *Empathy: Development, training, and consequences*. Hillsdale, N.J.: Erlbaum.

Gordon, D., S. Nowicki, and F. Wichern. 1981. Observed maternal and child behavior in a dependency-producing task as a function of children's locus of control orientation. *Merrill-Palmer Quarterly* 27:43–51.

Hoffman, M. L. 1960. Power assertion by the parent and its impact on the child. *Child Development* 31:129–43.

Hoffman, M. L. 1970. Moral development. In *Carmichael's manual of child psychology*, edited by P. H. Mussen, vol. 2. New York: Wiley.

Hoffman, M. L. 1975. Moral internalization, parental power, and the nature of parent-child interaction. *Developmental Psychology* 11:228–39.

Hoffman, M. L. 1976. Empathy, role-taking, guilt and the development of altruistic motives. In *Moral development and behavior*, edited by T. Lickona. New York: Holt, Rinehart and Winston.

Hoffman, M. L. 1981. The role of the father in moral internalization. In *The role of the father in child development*, 2d ed., edited by M. E. Lamb. New York: Wiley.

Hoffman, M. L. 1982. Affective and cognitive processes in moral internalization. In *Social cognition and social behavior: Developmental perspectives*, edited by E. T. Higgins, D. N. Ruble, and W. W. Hartup. Cambridge: Cambridge University Press.

Hoffman, M. L., and H. D. Saltzstein. 1967. Parent discipline and the child's moral development. *Journal of Personality and Social Psychology* 5:45–57.

Johannesson, I. 1974. Aggressive behavior among schoolchildren related to maternal practices in early childhood. In *Determinants and origins of aggressive behavior*, edited by J. deWitt and W. W. Hartup. The Hague: Mouton.

Lamb, M. E. 1977. Father-infant and mother-infant interaction in the first year of life. *Child Development* 48:167–81.

Lamb, M. E. 1981. Fathers and child development: An integrative overview. In *The role of the father in child development*, 2d ed., edited by M. E. Lamb. New York: Wiley.

Lefkowitz, M. M., L. D. Eron, L. O. Walder, and L. R. Huesmann. 1977. *Growing up to be violent*. New York: Pergamon.

Lewin, K., R. Lippitt, and R. White. 1939. Patterns of aggressive behavior in experimentally created social climates. *Journal of Social Psychology* 10:271–99.

Lewis, C. C. 1981. The effects of parental firm control: A reinterpretation of findings. *Psychological Bulletin* 90:547–63.

Loeb, R. C., L. Horst, and P. J. Horton. 1980. Family interaction patterns associated with self-esteem in preadolescent girls and boys. *Merrill-Palmer Quarterly* 26:203–17.

Maccoby, E. E., and J. A. Martin. 1983. Socialization in the context of the family: Parent-child interaction. In *Handbook of child psychology* (formerly *Carmichael's manual of child psychology*), 4th ed., edited by P. H. Mussen, vol. 4: *Socialization, personality, and social development*, edited by E. M. Hetherington, 1–101. New York: Wiley.

Patterson, G. R. 1976. The aggressive child: Victim and architect of a coercive system. In *Behavior modification and families.* Vol. 1: *Theory and research*, edited by L. A. Hamerlynck, L. C. Handy, and E. J. Mash. New York: Brunner-Mazell.

Patterson, G. R. 1979. A performance theory for coercive family interactions. In *The analysis of social interactions: Methods, issues and illustrations*, edited by R. B. Cairns. Hillsdale, N.J.: Erlbaum.

Patterson, G. R. 1980. Mothers: The unacknowledged victims. *Monograph of the Society for Research in Child Development* 45 (5, Serial N. 186).

Patterson, G. R. 1982. *Coercieve family process*. Eugene, Ore.: Castalia Press.

Patterson, G. R., and J. A. Cobb. 1971. A dyadic analysis of "aggressive" behavior. In *Minnesota symposium on child psychology*, vol. 5, edited by J. P. Hill. Minneapolis: University of Minnesota Press.

Pulkkinen, L. 1982. Self-control and continuity from childhood to adolescence. In *Life-span development and behavior*, edited by P. B. Baltes and O. G. Brim, vol. 4. New York: Academic Press.

Qadri, A. J., and G. A. Kaleem. 1971. Effect of parental attitudes on personality adjustment and self-esteem of children. *Behaviorometric* 1:19–24.

Saltzstein, H. D. 1976. Social influence and moral development: A perspective on the role of parents and peers. In *Moral development and behavior,* edited by T. Lickona. New York: Holt, Rinehart and Winston.

Sears, R. R. 1961. Relation of early socialization experiences to aggression in middle childhood. *Journal of Abnormal and Social Psychology* 63:466–92.

Yarrow, M. R., J. D. Campbell, and R. Burton. 1968. *Child rearing: An inquiry into research and methods.* San Francisco: Jossey-Bass.

Zahn-Waxler, C., E. M. Cummings, and R. Iannotti, eds. 1986. *Altruism and aggression: Biological and social origins.* Cambridge: Cambridge University Press.

Zahn-Waxler, C., M. Radke-Yarrow, and R. A. King. 1979. Child-rearing and children's prosocial initiations toward victims of distress. *Child Development* 50:319–30.

B3. Fundamentalist Christian Childrearing Manuals

Dobson 1970/1992 is the classic and probably the most moderate. Hyles is the most extreme.

Christenson, L. 1970. *The Christian family.* Minneapolis: Bethany House.

Dobson, J. 1970. *Dare to discipline.* Wheaton: Living Books/Tyndale House.

Dobson, J. 1978. *The strong-willed child: Birth through adolescence.* Wheaton: Living Books/Tyndale House.

Dobson, J. 1987. *Parenting isn't for cowards.* Dallas: Word.

Dobson, J. 1992. *The new dare to discipline.* Wheaton: Tyndale House.

Dobson, J., and Bauer, G. 1990. *Children at risk.* Dallas: Word.

Fugate, J. R. 1980. *What the Bible says about . . . child training.* Tempe: Alpha Omega.

Hyles, J. 1972. *How to rear children.* Hammond: Hyles-Anderson.

LaHaye, B. 1977. *How to develop your child's temperament.* Eugene, Ore.: Harvest House.

LaHaye, B. 1990. *Who will save our children?* Brentwood, Tenn.: Wolgemuth and Hyatt.

Swindoll, C. 1991. *The strong family.* Portland: Multnomah.

Tomczak, L. 1982. *God, the rod, and your child's bod: The art of loving correction for Christian parents.* Old Tappan: Fleming H. Revell.

B4. Mainstream Childrearing Manuals

Spock and Rothenberg is the updated version of the classic. The Brazelton books are currently extremely popular.

Bettelheim, B. 1987. *A good enough parent: A book on child-rearing.* New York: Alfred A. Knopf.

Brazelton, T. B. 1983. *Infants and mothers.* New York: Delacorte Press/Lawrence.

Brazelton, T. B. 1984. *Neonatal behavioral assessment scale.* 2d ed. Philadelphia: Lippincott.

Brazelton, T. B. 1984. *To listen to a child.* Reading: Addison-Wesley/Lawrence.

Brazelton, T. B. 1985. *Working and caring.* Reading: Addison-Wesley/Lawrence.

Brazelton, T. B. 1989. *Toddlers and parents.* Rev. ed. New York: Delacorte Press/Lawrence.

Brazelton, T. B. 1992. *On becoming a family.* Rev. ed. New York: Delacorte Press/Lawrence.

Brazelton, T. B. 1992. *Touchpoints.* Addison-Wesley.

Brazelton, T. B., and B. G. Cramer. 1990. *The earliest relationship.* Reading: Addison-Wesley/Lawrence.

Cramer, B. G. 1992. *The importance of being baby.* Reading, MA: Addison-Wesley/Lawrence.

Dreikurs, R. 1958. *The challenge of parenthood.* Rev. ed. New York: Hawthorn Books.

Dreikurs, R. 1964. *Children: The challenge.* New York: Penguin.

Fraiberg, S. 1959. *The magic years: Understanding and handling the problems of early childhood.* New York: Charles Scribner's Sons.

Ginott, H. G. 1956. *Between parent and child.* New York: Macmillan.

Gordon, T. 1975. *P.E.T.: Parent Effectiveness Training: The tested way to raise responsible children.* New York: New American Library.

Kimball, G. 1988. *50/50 parenting: Sharing family rewards and responsibilities.* Lexington: Lexington Books.

Leach, P. 1984. *Your growing child from babyhood through adolescence.* New York: Alfred A. Knopf.

Leach, P. 1989. *Your baby and child: From birth to age five.* Rev. ed. New York: Alfred A. Knopf.

Nelsen, J. 1981. *Positive discipline*. Fair Oaks, Calif.: Sunrise Press.

Nelson, J., L. Lott, and H. S. Glenn. 1993. *Positive discipline A-Z: 1001 solutions to everyday parenting problems*. Rockland, Calif.: Prima Publishing.

Popkin, M. 1987. *Active parenting: Teaching cooperation, courage, and responsibility*. San Francisco: Perennial Library.

Rosen, M. 1987. *Stepfathering*. New York: Ballantine Books.

Samalin, N., with M. Moraghan Jablow. 1987. *Loving your child is not enough: Positive discipline that works*. New York: Penguin Books.

Spock, B. 1988. *Dr. Spock on parenting: Sensible advice from America's most trusted child care expert*. New York: Simon and Schuster.

Spock, B., and M. B. Rothenberg. 1992. *Dr. Spock's baby and child care*. New York: Simon and Schuster.

Winnicott, D. W. 1987. *The child, the family and the outside world*. Introduction by M. H. Klaus. Reading: Addison-Wesley/Lawrence.

Winnicott, D. W. 1988. *Babies and their mothers*. Introduction by B. Spock. Reading: Addison-Wesley/Lawrence.

Winnicott, D. W. 1993. *Talking to parents*. Introduction by T. B. Brazelton. Reading: Addison-Wesley/Lawrence.

B5. Childrearing and Violence

Greven's book provides an overview of the research.

Altemeyer, B. 1988. *Enemies of freedom: Understanding right-wing authoritarianism*. San Francisco: Jossey-Bass.

Bandura, A. 1973. *Aggression: A social learning analysis*. Englewood Cliffs: Prentice-Hall.

Bruce, D. 1979. *Violence and culture in the antebellum South*. Austin: University of Texas Press.

Dobash, R. E., and R. Dobash. 1979. *Violence against wives: A case against the patriarchy*. New York: The Free Press.

Gelles, R. J. 1987. *The violent home*. Rev. ed. Newbury Park: Sage Publications.

Gelles, R. J., and M. A. Straus. 1988. *Intimate violence*. New York: Simon and Schuster.

Gil, D. G. 1970. *Violence against children: Physical abuse in the United States.* Cambridge: Harvard University Press.

Gordon, L. 1988. *Heroes of their own lives: The politics and history of family violence—Boston 1880–1960.* New York: Viking.

Greven, P. 1991. *Spare the child: The religious roots of punishment and the psychological impact of physical abuse.* New York: Alfred A. Knopf.

Huesmann, L. R., L. D. Eron, M. N. Lefkowitz, and L. O. Walder. 1984. Stability of aggression over time and generations. *Developmental Psychology* 20:1120–34.

Kelman, H. C., and V. L. Hamilton. 1989. *Crimes of obedience: Toward a social psychology of authority and responsibility.* New Haven: Yale University Press.

Lefkowitz, M. M., L. D. Eron, L. O. Walder, and L. R. Huesmann. 1977. *Growing up to be violent: A longitudinal study of the development of aggression.* New York: Pergamon Press.

Pagelow, M. D., and L. W. Pagelow. 1984. *Family violence.* New York: Praeger.

Pizzey, E. 1977. *Scream quietly or the neighbors will hear.* Hillside, N.J.: Enslow.

Pizzey, E., and J. Shapiro. 1982. *Prone to violence.* Feltham, England: Hamlyn Paperbacks.

Pleck, E. 1987. *Domestic tyranny: The making of social policy against family violence from colonial times to the present.* New York: Oxford University Press.

Renvoize, J. 1978. *Web of violence: A study of family violence.* London: Routledge and Kegan Paul.

Shupe, A., W. A. Stacy, and L. R. Hazlewood. 1987. *Violent men, violent couples: The dynamics of domestic violence.* Lexington: Lexington Books.

Straus, M. A., R. J. Gelles, and S. K. Steinmetz. 1981. *Behind closed doors: Violence in the American family.* Garden City: Anchor Books.

Taves, A., ed. 1989. *Religion and domestic violence in early New England: The memoirs of Abigail Abbot Bailey.* Bloomington: Indiana University Press.

Taylor, L., and A. Maurer. 1985. *Think twice: The medical effects of corporal punishment.* Berkeley: Generation Books.

Walker, L. E. 1984. *The battered woman syndrome.* New York: Springer.

B6. Critiques of Fundamentalist Childrearing

Bartkowski and Ellison is an excellent comparison of mainstream and fundamentalist approaches, as well as a rich source of research material.

Bartkowski, J. P., and Ellison, C. G. 1995. Divergent models of childrearing in popular manuals: Conservative Protestants vs. the mainstream experts. *Sociology of Religion* 56:21–34.

Boone, K. C. 1989. *The Bible tells them so: The discourse of Protestant fundamentalism.* Albany: SUNY Press.

Ellison, C. G., and J. P. Bartkowski. 1995. Religion and the legitimation of violence: The case of conservative Protestantism and corporal punishment. In *The web of violence: From interpersonal to global,* edited by L. R. Kurtz and J. Turpin. Urbana: University of Illinois Press.

Ellison, C. G., and D. E. Shertak. 1993a. Conservative Protestantism and support for corporal punishment. *American Sociological Review* 58:131–44.

Ellison, C. G., and D. E. Shertak. 1993b. Obedience and autonomy: Religion and parental values reconsidered. *Journal for the Scientific Study of Religion* 32:313–29.

Elshtain, J. B. 1990. The family in political thought: Democratic politics and the question of authority. In *Fashioning family theory,* edited by J. Sprey, 51–66. London: Sage.

Fromm, E. 1941. *Escape from freedom.* New York: Holt, Rinehart and Winston.

McNamara, P. H. 1985. Conservative Christian families and their moral world: Some reflections for sociologists. *Sociological Analysis* 46:93–99.

Nock, S. L. 1988. The family and hierarchy. *Journal of Marriage and the Family* 50:957–66.

Roof, W. C., and W. McKinney. 1987. *American mainline religion.* New Brunswick: Rutgers University Press.

Rose, S. D. 1988. *Keeping them out of the hands of Satan: Evangelical schooling in America.* New York: Routledge, Chapman, and Hall.

Wald, K. D., D. E. Owen, and S. S. Hill. 1989. Habits of the mind? The problem of authority in the New Christian Right. In *Religion and behavior in the United States,* edited by T. G. Jelen, 93–108. New York: Praeger.

Warner, R. S. 1979. Theoretical barriers to the understanding of evangelical Christianity. *Sociological Analysis* 40:1–9.

B7. Background: Childrearing and National Character

Current research on the effects of childrearing has an antecedent in earlier, very intensive research on the relationship between childrearing and national character. The best survey is Inkeles and Levinson. An important part of the history of the movement is discussed in Andrew Lakoff's paper on Margaret Mead's involvement.

Adorno, T. W., E. Frenkel-Brunswik, D. J. Levinson, and R. N. Sanford. 1950. *The authoritarian personality.* New York: Harper and Row.

Bateson, G., and M. Mead. 1942. *Balinese character: A photographic analysis.* New York: New York Academy of Sciences.

Benedict, R. F. 1946. *The chrysanthemum and the sword.* Boston: Houghton Mifflin.

Benedict, R. F. 1949. Child rearing in certain European cultures. *American Journal of Orthopsychiatry* 19:342–50.

Gorer, G. 1950. The concept of national character. *Science News* 18:105–23. Harmondsworth, England: Penguin Books.

Gorer, G., and J. Rickman. 1949. *The people of Great Russia.* London: Cresset Press.

Haring, D. G., ed. 1948. *Personal character and cultural milieu.* Syracuse: Syracuse University Press.

Inkeles, A., and D. J. Levinson. 1954. National character: The study of modal personality and sociocultural systems. In *Handbook of social psychology,* vol. 2: *Special fields and applications,* edited by G. Lindzey, chap. 26. Cambridge: Addison-Wesley.

Kardiner, A. 1945. The concept of basic personality structure as an operational tool in the social sciences. In *The science of man in the world crisis,* edited by R. Linton, 107–22. New York: Columbia University Press.

Lakoff, A. 1995. Margaret Mead's diagnostic photography. *Visual Anthropology Review,* Spring 1996.

Mead, M. 1951a. *Soviet attitudes toward authority.* New York: McGraw-Hill.

Mead, M. 1951b. The study of national character. In *The policy sciences,* edited by D. Lerner and H. D. Lasswell, 70–85. Stanford: Stanford University Press.

Mead, M. 1953. National character. In *Anthropology today,* edited by A. L. Kroeber, 642–67. Chicago: University of Chicago Press.

Whiting, J. W. M., and I. L. Child. 1953. *Child training and personality.* New Haven: Yale University Press.

C. Politics

C1. Conservative Political Writings

This is the tip of the iceberg—just the books mentioned in the text.

Bennett, W. J. 1992. *The de-valuing of America: The fight for our culture and our children.* New York: Simon and Schuster.

Bennett, W. J. (ed. with commentary). 1993. *The book of virtues: A treasury of great moral stories.* New York: Simon and Schuster.

Gillespie, E., and B. Schellhas. 1994. *Contract with America: The bold plan by Rep. Newt Gingrich, Rep. Dick Armey and the House Republicans to change the nation.* New York: Times Books/ Random House.

Gingrich, N. 1995. *To renew America.* New York: HarperCollins.

Limbaugh, R. 1993. *See, I told you so.* New York: Simon and Schuster.

C2. Neoconservatism

These are the writings of some prominent conservative intellectuals, some of whom started out further to the left.

DeMuth, C., and W. Kristol. 1995. *The neoconservative imagination: Essays in honor of Irving Kristol.* Washington, D.C.: AEI Press.

Ehrman, J. 1995. *The rise of conservatism: Intellectuals in foreign affairs, 1945–1994.* New Haven: Yale University Press.

Fukuyama, R. 1992. *The end of history and the last man.* New York: The Free Press.

Glazer, N. 1976. American values and American foreign policy. *Commentary,* July.

Kirkpatrick, J. 1988. Welfare state conservatism: Interview by Adam Meyerson. *Policy Review,* Spring.

Kristol, I. 1976. What is a "neo-conservative"? *Newsweek*, January 19, p. 17.

Kristol, I. 1983. *Reflections of a neoconservative: Looking back, looking ahead*. New York: Basic Books.

Kristol, W. 1993. A conservative looks at liberalism. *Commentary* 96(3):33–36.

Kristol, W. 1994. William Kristol looks at the future of the GOP. *Policy Review* 67:14–18.

Lipset, S. M. 1988. Neoconservatism: Myth and reality. *Society*.

Moynihan, D. P. 1993. Defining deviancy down. *American Scholar*, Winter.

Moynihan, D. P. 1993. Toward a new intolerance. *Public Interest*, Summer.

Muravchik, J. 1991. *Exporting democracy*. Washington: American Enterprise Institute.

Podhoretz, N. 1979. *Breaking ranks*. New York: Harper and Row.

Sowell, T. 1987. *A conflict of visions: Ideological origins of political struggles*. New York: William Morrow.

Wilson, J. Q. 1980. Neoconservatism: Pro and con. *Partisan Review* 4.

Wilson, J. Q. 1993. *The moral sense*. New York: The Free Press.

C3. Modern Theoretical Liberalism

John Rawls's *A Theory of Justice* is the most influential work in modern theoretical liberalism, which seeks to consider social issues such as poverty, health, and education in the same arena as individual rights.

Arneson, R. 1989. Introduction. A symposium on Rawls's *Theory of Justice:* Recent developments. *Ethics* 99:695–710.

Daniels, N. 1978. *Reading Rawls: Critical studies of* A Theory of Justice. Oxford: Basil Blackwell.

Harsanyi, J. 1976. *Essays on ethics, social behaviour and scientific explanation*. Dordrecht: Reidel.

Kukathas, C., and P. Pettit. 1990. *Rawls:* A Theory of Justice *and its critics*. Stanford: Stanford University Press.

Mulhall, S., and A. Swift. 1992. Liberals and communitarians. Cambridge, Mass.: Basil Blackwell.

Pogge, T. W. 1989. *Realizing Rawls*. Ithaca: Cornell University Press.

Rawls, J. 1971. *A theory of justice*. Cambridge: Belknap Press/Harvard University Press.

Rawls, J. 1982. The basic liberties and their priority. In *The Tanner Lectures on Human Values*, edited by S. MacMurrin, 3:1–89. Cambridge: Cambridge University Press.

Rawls, J. 1993. *Political liberalism*. New York: Columbia University Press.

Raz, J. 1986. *The morality of freedom*. Oxford: Oxford University Press.

Rorty, R. 1982. *Consequences of pragmatism: Essays: 1972–1980*. Brighton: Harvester Press.

C4. Communitarian Critiques

Theoretical liberalism focuses on the individual and individual rights. Communitarian critiques claim that it makes no sense to think of individuals as separate from their communities and that responsibilites must be considered alongside rights.

Bellah, R., et al. 1985. *Habits of the heart: individualism and commitment in American life*. Berkeley: University of California Press.

Daly, M., ed. 1994. *Communitarianism: A new public ethics*. Belmont: Wadsworth.

Etzioni, A. 1988a. *The moral dimension: Toward a new economics*. New York: The Free Press.

Etzioni, A. 1988b. *The spirit of community: Rights, responsibilities, and the communitarian agenda*. New York: Crown.

Etzioni, A., ed. 1995. *New communitarian thinking: Persons, virtues, institutions, and communities*. Charlottesville: University of Virginia Press.

Gutmann, A. 1985. Communitarian critics of liberalism. *Philosophy and Public Affairs* 14:308–22.

Kukathas, C., and P. Pettit. 1990. The communitarian critique. In *Rawls: A Theory of Justice and its critics*, edited by C. Kukathas and P. Pettit, 92–118. Stanford: Stanford University Press.

MacIntyre, A. 1986. *After virtue: A study in moral theory*. 2d ed. London: Duckworth.

Sandel, M. 1982. *Liberalism and the limits of justice*. Cambridge: Cambridge University Press.

Spragens, T. A., Jr. 1995. Communitarian liberalism. In *New communitarian thinking,* edited by A. Etzioni, 37–51. Charlottesville: University of Virginia Press.

Taylor, C. 1985. Atomism. In his *Philosophical Papers,* vol. 2:187–210. Cambridge: Cambridge University Press.

Walzer, M. 1981. Philosophy and democracy. *Political Theory* 9: 379–99.

Walzer, M. 1983. *Spheres of justice*. Oxford: Basil Blackwell.

Wolff, R. P. 1977. *Understanding Rawls: A reconstruction and critique of* A theory of justice. Princeton: Princeton University Press.

C5. Theoretical Libertarianism

Theoretical libertarians maintain a pure focus on individual rights.

Nozick, R. 1974. *Anarchy, state and utopia*. New York: Basic Books.

D. Public Administration

D1. Bureaucratic Reform

Osborne and Gaebler is the classic, providing a blueprint for the Clinton administration's attempted reform of the bureaucracy. Barzelay documents an example of bureaucratic reform in the state of Minnesota.

Barzelay, M., with B. J. Armajani. 1992. *Breaking through bureaucracy: A new vision for managing in government*. Berkeley: University of California Press.

Drucker, P. 1985. *Innovation and entrepreneurship*. New York: Harper and Row.

Drucker, P. 1989. *The new realities*. New York: Harper and Row.

Kanter, R. M. 1983. *The change masters: Innovation and entrepreneurship in the American corporation*. New York: Harper and Row.

Osborne, D., and T. Gaebler. 1992. *Reinventing government: How the entrepreneurial spirit is transforming the public sector*. New York: Addison-Wesley.

Wilson, J. Q. 1989. *Bureaucracy: What government agencies do and why they do it*. New York: Basic Books.

D2. Star Wars Policy

This is a classic on why Star Wars isn't feasible. It has become relevant again in the current debate over the resurrection of the Star Wars program.

Lakoff, S., and H. F. York. 1989. *A shield in space? Technology, politics, and the Strategic Defense Initiative: How the Reagan administration set out to make nuclear weapons "impotent and obsolete" and succumbed to the fallacy of the last move*. Berkeley: University of California Press.

E. Miscellaneous

Lovejoy, A. O. 1936. *The great chain of being: A study of the history of an idea*. Cambridge: Harvard University Press.

Index